Globalizing On-line: Telecollaboration, Internationalization, and Social Justice

TELECOLLABORATION IN EDUCATION

Vol. 4

Edited by
Melinda Dooly & Robert O'Dowd

PETER LANG
Bern · Berlin · Bruxelles · Frankfurt am Main · New York · Oxford · Wien

Nataly Tcherepashenets (ed.)

Globalizing On-line: Telecollaboration, Internationalization, and Social Justice

PETER LANG
Bern · Berlin · Bruxelles · Frankfurt am Main · New York · Oxford · Wien

Bibliographic information published by die Deutsche Nationalbibliothek
Die Deutsche Nationalbibliothek lists this publication in the Deutsche Nationalbibliografie; detailed bibliographic data is available on the Internet at ‹http://dnb.d-nb.de›.

British Library Cataloguing-in-Publication Data: A catalogue record for this book is available from The British Library, Great Britain

Library of Congress Control Number: 2014958382

ISBN 978-3-0343-1520-3 pb.
ISSN 1662-3037 pb.

ISBN 978-3-0351-0810-1 eBook
ISSN 2235-7599 eBook

This publication has been peer reviewed.

Hochfeldstrasse 32, CH-3012 Bern, Switzerland
info@peterlang.com, www.peterlang.com

To Michael

May you live your life inspired by a passion for learning and by a concern to strive for social justice.

Contents

Series Editors' Preface 9

Acknowledgements 11

Notes on Contributors 13

NATALY TCHEREPASHENETS
Introduction 19

Part I: Telecollaboration as a Practice of Inclusion

HAYAT AL-KHATIB
Globalizing Online: Binary Distinctions
or Global Partaking? 35

ANNA TURULA
Telecollaboration in the Clash of High-
and Low-context Cultures: Social Justice at Stake 55

REBEKAH MCPHERSON, KELLEY KING, ANDREW MILSON,
AND MARY BAILEY ESTES
Evaluation of an International Online Learning
Initiative in Special Education 85

Part II: Web 2.0 and Social Change

DERYA KULAVUZ-ONAL AND CAMILLA VÁSQUEZ
Opening up Worlds: Intercultural Exchanges
through Telecollaboration 109

DOMINIQUE VINET
Online collaboration with the language-learning
software @genda 2.0 129

REBECCA CHARRY ROJE AND DAVID S. MARTINS
Between "Pleasantville" and "My Way or the Highway": Promoting Productive Discussion of Social Justice in a Globally Linked Learning Environment 151

Part III: Intercultural Encounters as Transformative Experiences

WENDY ANDERSON AND JOHN CORBETT
'What Do We Chat about When We Chat about Culture?' The Discourse of Online Intercultural Exchanges 177

MARINA ORSINI-JONES, ELWYN LLOYD, ZOE GAZELEY, GWENOLA BESCOND AND BEATRIZ VERA LÓPEZ
Student-driven intercultural awareness raising with MexCo: agency, autonomy and threshold concepts in a telecollaborative project between the UK and Mexico 201

HIMANEE GUPTA-CARLSON
Internationalizing U.S. Students with Hip-hop and Social Networks 235

NATALY TCHEREPASHENETS
Conclusion 257

Index 263

Series Editors' Preface

This series is dedicated to promoting a wider understanding of the activity of telecollaboration in educational settings. Telecollaboration refers to the engagement of groups of students in online intercultural interaction and collaboration with partner classes from other cultural contexts or geographical locations, under the guidance of educators and/or expert facilitators. The application of such activity may include different subject areas (e.g. Foreign Language Education, History, Science) as well as different educational contexts, including but not limited to primary, secondary, university and adult education.

This fourth volume in our series looks at a topic that is fast becoming one of the most prevalent concerns around the world: social justice in education, and in particular how telecollaborative exchange can play a role in experiential learning of key features related to this concept. Inevitably the concept itself is fraught with controversy related to two key questions: What do we expect education to do? What is its purpose?

> These may seem naïve questions, but they lie at the heart of the debate about the nature of inequity in our societies and the future development of social activity. There is no simple and universally agreed answer: different societies have debated and disputed the reasons for why we should teach and learn [since the times of Aristotle]. Do we teach to enhance and develop the individual's intelligence or their social behaviour? Do we spend large sums of money on schooling in order to develop a skilled and able workforce ('the useful things in life'), or to advance socially responsible behaviour ('those most conducive to virtue'), or to support specialised knowledge and progression ('exceptional accomplishments')? Are these competing alternatives, or can some or all be achieved at the same time? (Ross, Dooly & Hartsmar, 2012, p. 4)

Just as the basis for social inequity and educational disadvantage is not a straightforward, transparent cause-and-effect process, the discussion of social justice in education is, at times, circular and complex. This is reflected in the following chapters as they approach the topic from different perspectives, although various recurrent themes do emerge as the authors describe how the development of social justice stands to benefit from the use of telecollaboration. These themes include the opportunities online intercultural exchange offers for collaboration, critical thinking and reflection on controversial topics and intercultural understanding.

Telecollaboration has now been in use as a pedagogical tool in foreign language education for almost 25 years. While the first reports of online exchange were primarily practical publications describing how the activity was being carried out (Warschauer, 1995), recent years have seen a proliferation of research publications which have explored the learning outcomes of telecollaborative initiatives (see, for example, Belz & Thorne, 2006; Guth & Helm, 2010). However, these studies have looked primarily at the potential benefits of telecollaboration for specific aspects of language learning, including students' linguistic competence, aspects of intercultural competence, autonomy and, more recently, digital literacies. The complex issue of social justice and how telecollaboration can support its development in institutionalised learning contexts, has rarely been explicitly addressed until now.

This fourth volume therefore provides timely material for educators, researchers and policy-makers interested in understanding how telecollaboration can contribute to the creation of a more equitable, respectful, and just society in an attempt to go beyond mere hollow rhetoric.

Melinda Dooly, Universitat Autònoma de Barcelona, Spain
Robert O'Dowd, Universidad de León, Spain

4 November 2014

References

Belz, J., & Thorne, S. (Eds) (2006). *Internet-mediated intercultural foreign language education*. Boston, MA: Heinle & Heinle.

Guth, S., & Helm, F. (Eds.) (2010). *Telecollaboration 2.0: Language, literacies and intercultural learning in the 21st Century*. Bern: Peter Lang.

Ross, A., Dooly, M. & Hartsmar, N. (2012). *Equalities and education in Europe: Explanations and excuses for inequality.* Newcastle upon Tyne: Cambridge Scholars Publishing.

Warschauer, M. (Ed.) (1995). *Virtual connections: Online activities and projects for networking language learners.* Honolulu, HI: University of Hawai'i Second Language Teaching and Curriculum Center.

Acknowledgements

I would like to express my gratitude to the contributors to this volume for their innovation, knowledge, and insight. I would also like to thank Melinda Dooly, Robert O'Dowd and Benjamin Fröhlich at Peter Lang for their editorial guidance and encouragement.

During my work on this volume, I was given several opportunities to present and discuss my work at professional conferences. I am grateful to the dean, Dr. Tom Mackey and to the Office of Academic Affairs of the Empire State College, State University of New York, for their funding of this project, and for their generous support of my initiatives, which led to its completion. One of them was the opportunity to present at the fourteenth Cambridge International Conference, "Internationalization and Social Justice: the role of Open, Distance and e-Learning," organized by Anne Gaskell, Roger Mills and Alan Tait, where I had the pleasure to meet some of the future contributors to this volume. My work as a fellow at the State University of New York's Collaborative Online International Institute, as part of a National Endowment for the Humanities supported fellowship for teaching humanities across national borders, was another such opportunity.

Special thanks go to Dr. Florence Lojacono (Universidad de Las Palmas, Gran Canaria) for productive discussions during the early stages of preparing this manuscript. I would also like to thank Dr. Janet Shideler, Dean of School of Liberal Arts at Siena College, for her interest and feedback on the project from its inception. Dr. Miriam Russell is responsible for many other improvements in this book. A debt of gratitude is also owed to my family for their support, love and understanding.

Notes on Contributors

Hayat Al-Khatib is Associate Professor of Applied Linguistics at the Arab Open University – Lebanon. She was awarded a PhD in Applied Linguistics from The University of London in 2003. She has had ample professional experience in Britain with the Centre of Information on Language Teaching (CILT) in consultancy and advising, teaching at the Institute of Education – University of London and as research associate at Goldsmith – University of London. Al-Khatib is currently Head of the English department at the Arab Open University in Lebanon and the Academic Deputy Director. Hayat Al-Khatib has several publications in the fields of Applied Linguistics and English Language Teaching, Open Distance and E-Learning, Pragmatics, Discourse Analysis and Functional Grammar. She serves as associate editor in the *Linguistic Journal, the European Scientific Journal* (ESJ) and is the editor in chief of the *Centre of Applied Linguistics Research (CALR) journal*. Her e-mail is hkhatib@aou.edu.lb.

Wendy Anderson is Senior Lecturer in English Language at the University of Glasgow. Her research and teaching interests include intercultural language education, semantics, metaphor, corpus linguistics, and translation. She was formerly Research Associate for the AHRC-funded Scottish Corpus of Texts & Speech project (<www.scottishcorpus.ac.uk>). She is the co-author of *Exploring English with Online Corpora* (with John Corbett, Palgrave Macmillan, 2009), and the author of *The Phraseology of Administrative French* (Rodopi, 2006). She directs the AHRC-funded project, Mapping Metaphor with the *Historical Thesaurus* (<www.glasgow.ac.uk/metaphor>).

Gwenola Bescond is Course Leader for the French and English degrees in the department of English and Languages at Coventry University (UK). She is currently working on her PhD on a comparative study between France and England, investigating barriers and enablers to inclusion within Higher Education for students with disabilities – in particular those with visual impairment. Her research interests include social justice, international approaches to education, international and national inclusion discourse, support for learning, Computer Assisted Languages Learning

(CALL) and translation. Her MA considered inequalities regarding Dalit women's education in Tamil-Nadu. She has created blended learning materials for Business French and is currently contributing to the MexCO CMC project.

REBECCA CHARRY ROJE is Senior Lecturer in English at Rochester Institute of Technology's Dubrovnik, Croatia, campus, where she has taught English language and literature to international students since 2001. Her recent articles include "Communicative language teaching in Croatian English classrooms" published by the European Council on Hotel, Restaurant & Institutional Education, and she is co-author of a forthcoming English language textbook for use in Croatian college preparatory high schools. She works frequently as an editor and translator, and holds degrees in English Literature and Writing from Haverford College and Georgetown University.

JOHN CORBETT is a Professor of English at the University of Macau, and an Honorary Senior Research Fellow in English Language at the University of Glasgow. He has published widely on intercultural language education and on corpus linguistics. He was Principal Investigator of the AHRC-funded Scottish Corpus of Texts & Speech project, and the Corpus of Modern Scottish Writing (1700–1945) (<www.scottishcorpus.ac.uk>). He is the author of *An Intercultural Approach to English Language Teaching* (Multilingual Matters, 2003) and *Intercultural Language Activities* (Cambridge University Press, 2011) and he co-authored *Exploring English with Online Corpora* (with Wendy Anderson, Palgrave Macmillan, 2009). In 2013, he was Visiting Professor in English at the University of Sao Paulo in Brazil.

ZOE GAZELEY is a lecturer in English for Academic Purposes and a research assistant for the MexCo project, in the Department of English and Languages at Coventry University (UK). Zoe has previously taught Business English in Mexico City and has also taught EFL and EAP in several other countries. Her research interests include intercultural communication in language learning and teaching along with the development of digital literacies in language learning. Her MA in English Language Teaching was based on the pilot phase of a CMC international project with Mexico.

HIMANEE GUPTA-CARLSON is Assistant Professor at SUNY Empire State College, where she teaches hip-hop history, storytelling, and social justice. Born in the Midwest to immigrant Indian parents, she writes about global diasporas and their impact on Middle America in her forthcoming book *Muncie, India(na)*. Her contribution to this volume resulted from a course she created with University of Victoria Assistant Professor Janni Aragon as part of a National Endowment for the Humanities supported fellowship on teaching humanities across national borders. She also has worked with other hip-hop scholars on a document that establishes Hip-Hop Studies as an educational endeavor dedicated to promoting social justice through connections between American inner cities and urban environments globally.

Kelley M. King is an Associate Professor of Curriculum and Instruction at the University of North Texas. Her research interests include the history of education, civic education, and the lives of women educators. She is the author of the monograph, *Call her a citizen: Progressive-era activist and educator Anna Pennybacker*, and is currently conducting research on a history of the Texas State Board of Education.

Derya Kulavuz-Onal has recently earned her PhD from the University of South Florida and joined the Department of English at Salisbury University, Maryland, as a new faculty member in Fall 2013. She taught English in Turkey and in the USA for nine years prior to working with pre-service language teachers. Her academic interests intersect on the broader fields of language teacher education, applied linguistics, instructional technology in language teaching, and (online) qualitative research. She can be reached at <kulavuzd@gmail.com>; and her dissertation-related work can be found at <https://sites.google.com/site/wianetnography/>.

Elwyn Lloyd is Course Director for TEFL (Teaching English as a Foreign Language) in the Department of English and Languages at Coventry University (UK). Previous to his current role he had accumulated extensive experience as an English teacher and Director of Studies at language schools in Spain, and he obtained his MA in Language Learning and Technology. His research interests are centred on the application of telecollaboration, social networking sites, and mobile technology in language learning and teaching. His most recent publications relate to the use of Web 2.0 tools for language learning and teaching.

Beatriz Vera López is Associate Professor of English in the department of foreign languages at FES-Z (Facultad de Estudios Superiores Zaragoza) UNAM (Universidad Nacional Autónoma de México). In 2012 she completed a PhD at the University of Nottingham UK where she discussed how the study of Modern Languages and Cultures should include the 'agentification' of the learner to enable them to make sense of their experience in a foreign language and culture. Her MA, also in the University of Nottingham, was on the role of the poetic features of language in SLA and the language learner autonomy. She has designed blended materials for EFL and published work on multilingual intercultural issues.

David S. Martins is Associate Professor and founding director of the University Writing Program at Rochester Institute of Technology. He collaborated with Rebecca Roja Charry during their Fellowship with the SUNY Collaborative Online International Learning (COIL) Institute. His work has won the Best Article of the Year 2008 in *Teaching English in the Two Year College*, and has appeared in *Communication Studies* andthe *Journal of Medical Humanities*. His digital essay "Transnational Writing Programs and Emerging Models for Writing, Teaching, and Learning" is forthcoming in *CCC-Online*, and his edited collection, *Transnational Writing Program Administration* will be published by Utah State University Press.

Andrew Milson is Professor of History and Geography at the University of Texas at Arlington. His current teaching and scholarship focus on historical geography and the history of education. He is the co-editor of four books and has published articles in journals such as Theory and Research in Social Education, Educational Policy, and Journal of Geography.

Marina Orsini-Jones is Associate Head in the Department of English and Languages at Coventry University (UK) and leads the Applied Research Group Pedagogical Innovation in Languages and Literature. She has been engaged with CALL (Computer Assisted Language Learning) since writing her MA dissertation on this topic in 1986. She has published student-centred multimedia courseware (*La neve nel bicchiere*, 1998) and scholarly work on action-research-led curricular innovation, digital literacies and e-learning. She has presented papers (including keynotes) at various national and international conferences and led national and international teams for e-learning projects.

Laura Pibworth is a lecturer in Italian and English and a research assistant for the MexCo project in the Department of English and Languages at

Coventry University (UK). Her research interests include Computer Assisted Language Learning (CALL) and Computer Mediated Communication (CMC) through Social Networking Sites. She has published work on SNSs for language learning based on her MA dissertation and co-edited conference proceedings.

REBEKAH MCPHERSON is Core Faculty in the Digital Teaching and Learning Department at The American College of Education. Her areas of interest include online education and digital tools for the classroom. She co-authored the article Perceptions of Special Education Pre-Service and In-service Teachers using a Simulated Classroom in the *Journal of Technology in Teacher Education* (2011). She presents at several technology conferences a year and co-developed a MOOC called, *Digital Tools for the K-12 Educator*, hosted by Canvas.net.

NATALY TCHEREPASHENETS is Associate Professor of Spanish and Area Coordinator of World Languages at the State University of New York, Empire State College. She received her PhD in Hispanic Languages and Literatures from the University of California, Los Angeles. Her research interests include such topics as intercultural language education and its impact on identity development, international collaborations and online learning cultures. She is also the author of *Place and Displacement in the Narrative World of Jorge Luis Borges and Julio Cortázar* (2008), a book chapter (with L. Snyder) "Transformations: Lifelong learners in the era of globalization" (in *Lifelong Learning and Social Justice*, edited by S. Jackson, NIACE 2011) and numerous articles on topics of teaching languages and cultures online, Latin American, Peninsular, and comparative literature, as well translation studies.

ANNA TURULA is an experienced EFL teacher and teacher trainer. Her particular interests lie in e-teacher training via the Moodle and social software (wikis). At present, she is Deputy Head of the School of English and a teacher trainer with the University of Social Sciences, Warsaw, Poland. She has published three books and numerous articles as well as co-edited two monographs. Her book *Form-Focused Instruction and the Advanced Language Learner. On the Importance of the Semantics of Grammar* was published by Peter Lang in 2011. Her ICT publications include, among others: 'Classroom dynamics of virtual learning environments: a humanistic perspective' (*CALL Review* 2010/2) as well as 'Micro- and macro-perspectives on students' attitudes to online classes' and 'Affect in VLEs: anxiety and motivation in blended EFL teacher training', both in press.

CAMILLA VÁSQUEZ is an Associate Professor of Applied Linguistics in the Department of World Languages at the University of South Florida, where she teaches graduate students in the MA-TESL program and in the interdisciplinary SLA/IT (Second Language Acquisition and Instructional Technology) doctoral program. Camilla's past research has examined issues in language teacher education, by focusing on the discourse of teacher/mentor interactions. More recently, Camilla's research interests have shifted to the study of digital texts and digital practices. Her forthcoming book, The Discourse of Online Reviews, will be published by Bloomsbury. Several of Camilla's publications can be accessed at <camillavasquez.com>.

DOMINIQUE VINET is a professor of English at the University of Bordeaux (France). He has specialized in online teaching and developed @genda 2.0, an e-learning platform dedicated to languages, and Erasmuslink, a software meant for the follow-up of students on mobility programs. He has also published a number of articles and books based on a psycho-analytical approach of modern British literature. He is presently in charge of international relations and student mobility, and head of the department of languages in the faculty of law and economics.

Nataly Tcherepashenets

Introduction

Action is the soul of the world
–Denis Diderot

Telecollaboration and Lessons in World Citizenship

A concern for social justice is the responsibility of educators and all citizens who cherish democracy and would like to contribute to long-lasting peace in a globalized world. This concern has inspired a variety of philosophical, political and sociological discourses and informed multiple debates in the field of education. Two interrelated themes stand out in these discussions: the right of individuals for self-realization and the quest for recognition. These themes are firmly linked to the relationship which thc Russian philosopher, Bakhtin (1993) viewed as intrinsic to the "architectonics of being," around which all values of life and culture are organized: "I, the other, and I-for-the–other" (p. 54). At this time of global cross-mingling of people, this relationship becomes increasingly complex, dynamic and unpredictable. Indeed, the nature of this relationship is at the core of our understanding of the concept of social justice.

Moving from abstract models to the world of action for the purpose of this book, we would like to define social justice as an educational practice. The main objective of this practice is to make it possible for motivated learners of different social, economic, cultural backgrounds to receive equal educational outcomes and to become responsible world citizens. We suggest that this can be achieved through access to rigorous academic content, collaboration based on mutual respect and open-mindedness, and teaching/design methods that enhance critical thinking, appeal to students' emotions, and encourage exploration of controversial topics.

Online intercultural exchanges open numerous opportunities for the performance of this educational practice, which acquires seminal significance at this time when the world paradoxically becomes both increasingly interdependent and strictly regionalized. In spite of the constant subversion of inside/outside opposition in "the space of flows" (Castells 1996: 406), created by the Internet, boundaries do not disappear. On the contrary, as Friedman (1999) perceptively observes, "they seem to be erected on every new street corner of every declining neighborhood of our world" (p. 14). Thus, globalization reveals one of the biggest challenges of our time: the exercise of the rules of mutual hospitality, which Kant in 1784 viewed as a basic human necessity.

Offering creative ways to achieve a shift from isolation to communication, telecollaboration can meet this challenge in addition to the development of common understanding of the benefits of cooperation and coexistence. This requires recognition of the essential humanity of others, an ability to question one's own previously formed points of view, and openness for a dialogue that is informed by mutual respect, tolerance and solidarity. We suggest that similar skills and competencies are intrinsic for the peaceful and enjoyable coexistence of world citizens.

Many distinguished scholars point out the strong correlation between the values nurtured by education and those which are useful in everyday life. Drawing a parallel between democracy and education, Dewey (1997) considers both institutions as "methods of value- creation" (p. 4). Viewing education as a political act, Freire (1998) states that "the posing of the problems of human beings in their relationship with the world" (p. 26) is among the major duties of education. Two relevant approaches, critical pedagogy and collaborative learning, emphasize that a key responsibility of education is to prepare an individual to participate in dialogue on equal terms in any setting, and for her/his mission as a producer of history and culture. In his pioneering work on the social and relational character of education, Vygotsky (1997) connects learning and activism, and associates the ability to collaborate and a belief in collaborative achievement with the capacity to know and transform one's world and become oneself.

Following these lines, and bridging collaborative learning, critical pedagogy and transformative practice, Stetsenko's (2008) concept of "collaborative purposeful transformation of the world" (p. 474) is intrinsically linked to the ideals of social justice and emancipation as the principle that grounds learning and development at all levels. Globalization, which de Sousa Santos (1999) views as "bundles of social relationships" (p. 215), shapes the interaction between the learner and the world today. This process, in turn, brings internationalization, which Knight and Altbach (2007) view as primarily policies and practices within higher education systems that are used to address the global environment, to the forefront of the agendas of education policy-makers as well as in the work of many researchers and practitioners.

Telecollaboration, which O'Dowd (2012) defines as the application of online communication tools with the purpose of bringing together classes of learners in geographically distant locations in order to enhance the development of content- related skills and intercultural competence through collaborative tasks and project work, exemplifies internationalization practices *par excellence*. Evoking dynamics and diversity, which globalization brings to everyone's lifestyle, telecollaboration (if approached appropriately) can encourage inclusive democratic education, and enhance tolerance and solidarity. As the chapters in this book demonstrate, this innovative learning/teaching approach, based on building knowledge and sharing experiences across the borders, can contribute positively to learners' intellectual growth as well as their education in world citizenship. Both processes are central to the life-learning in the 21st century.

Technology and Poiēsis of Learning

Online intercultural exchanges which allow perspective transformation and identity renewal through the "constructive engagement with otherness," (Daloz 2000: 110) exemplify learning opportunities at the intersections between internationalization and social justice. Offering innovative strategies to implement internationalization efforts, telecollaboration makes international learning increasingly accessible to students all over the world, because it does not require mobility, which often reproduces inequalities (Archer, 2007).This practice is inclusive for students who are geographically separated from dominant educational structures, and it opens multiple ways for interdisciplinary and inter-institutional collaborations. Though there are billions of people around the globe without Internet, as statistics illustrates, the digital divide is shrinking. For example, according to Amirault and Vasser (2010), the users of internet worldwide in 2010 grew 444. 8% in comparison with 2000 (<http://www.internetworldstats.com>). The authors notice another problem, which is an Internet penetration, which affects the access to educational resources. According to the Internet World Statistics in 2010 (June), penetration rates are 77.4% in North America, 58.4% Europe, 10.9% in Africa. The increased access to the internet may support internalization for the University.

Indeed, enhanced by technological advances, which emerge in response to social and cultural imperatives, telecollaboration mirrors the functioning of what Castells (2004) has called the "global society as a network", where participants seek opportunities for expression and connection. In order to succeed in this task, they need to practice "sociality," which Papacharissi (2011) aptly defines as the "sum of social behaviors that permit the individual to traverse from the state of individuality to that of sociality and fellowship" (p. 316). Furthering this insight, we suggest that this practice requires open communication, where the presence of one's stories, told in a way which respects and preserves 'the other' and by extension 'oneself,' is a key to success. In the context of telecollaboration, it is a necessary condition for the exercise of the ability and willingness of learners to contribute to collaborative knowledge development, which is instrumental for the empowerment of individuals.[1] This has been long perceived as one of the major goals of education. According to Freire (1998), for example, empowerment can be achieved by means of teaching/learning through "problem-posing," taking responsibility for one's own learning and realizing that everyone can be the "maker and dreamer of history," (p. 41) which is filled with possibilities. One may add that in today's connected world, genuine empowerment is associated with responsible decision- making, on both local and global scales, which requires the building and rebuilding of social relationships, based on open communication, in addition to honing of professional and personal skills.

At their best, intercultural exchanges can serve as a model of a hospitable and friendly setting for mutually enriching cooperation of learners from different parts of the world, working towards full development of

1 According to Shirky, individuals can take the full advantage of knowledge through the creation of "cultures of sharing" (2010: 143).

their self-esteem as they advance their potential and the proper use of their abilities.

This use of telecollaboration in education evokes Heidegger's view of technology as "unlocking, transforming, storing, distributing and switching about" (1993: 332), which leads to "unconcealment", that is "bringing –forth". The German philosopher associates this process with "a way of revealing" (p. 318) the 'truth' in his essay *The Question Concerning Technology* (1993). He defines this process of "unconcealment," which technology brings forward as "Enframing," (*Gestell),* which means "some kind of apparatus (for instance, a bookrack) as well as being the name for a skeleton" (1993: 325). Enframing in the sense of *poiēsis* lets "presences come forth into unconcealment" (1993: 326). Intriguingly, this process sounds familiar to the readers of Heidegger's (2001) "Origin of the Work of Art." Indeed, drawing a parallel between technology and art, the German philosopher emphasizes that both have to do with revealing. Further, Heidegger suggests that technology is "something *poiētic*," through which something is "produced."

Referring to the etymological analysis of the word, the German philosopher reminds readers that "Techne" defines the skills of the craftsman as well as the arts of the mind and the fine arts.[2] Thus, technology as both *physis* and *poiēsis*, according to Heidegger, challenges a person "forth into revealing" (1993: 326), which is a liberating act. As he perceptively points out, "when we open ourselves expressly to the essence of technology we find ourselves unexpectedly taken into a freeing claim" (1993: 330–331).

Online intercultural exchanges illustrate Heidegger's approach to technology *par excellence*. Opening up a space for discussion of most controversial topics, and challenging students and instructors alike into revealing and freeing themselves from prejudice and stereotypical thinking, telecollaboration brings into being the "enframed" 21st century episteme. This "revealing" is brought forth from concealment to unconcealment through the intersections of knowledge and art, integral to

2 Pelletier's (2004) approach to educational technology intriguingly coincides with Heidegger's rejection of the vision of technology as just a craft or technique. As this British scholar perceptively notices, "Educational technologies [...] are systems of cultural transmission, creating new contexts within which existing social interest express themselves" (p. 13). One of these contexts is a flow of information, which appears to undermine inside/outside opposition.

learner friendly instructional design and teaching, thereby inspiring mastery of the subject matter, critical thinking, life-long learning, tolerance and open-mindedness; all of which, as has been mentioned previously, are key factors to effective educational practices aimed at ensuring responsible world citizens who are aware of the importance of social justice and democracy. These key factors, combined with the students' willingness and ability to engage in open communication and collaborative knowledge creation, allows learning as *poiēsis* to happen, where *poiēsis* is both "a response to an overpowering experience of absence and instability" and a process which involves transformation "a moment of ecstasis when something moves away from its standing as one thing to become another" (Heidegger 1987: 32).

There are at least two obstacles to this process: global racism, which opens the door to the paralyzing power of stereotypes and prejudice, and exceptionalism, which leads to isolation and prospering ignorance.[3] Both challenges impede collaboration, undermine the possibility of dialogue on equal terms, and make one forget that people share the world and its problems.

The overcoming of these challenges, the first of which has been discussed in detail in earlier work by Tcherepashenets with Snyder (2011), requires a change in the perception of the world order as a set of binary oppositions, such as 'us' vs. 'them' or 'global' vs. 'local'. It also entails the questioning of views of society and culture as being defined exclusively by geographical borders and local communities. Telecollaboration, where every participant can be viewed as *hôte*, that is both the host and the guest, one that both gives and receives, offers and accepts, invites and is invited, can and should facilitate the overcoming of these challenges. By allowing learners from different parts of the globe to bring their cultures, struggles and histories from absence to presence through personal voices and experiences, online intercultural exchange often leads to the challenge of students' frame of reference that may result in the shift in their perceptions of each other and themselves (Dooly, 2008; Guth & Helm, 2012; O'Dowd, 2012). Telecollaboration can be also instrumental in discovering that human needs for autonomy, competence and relatedness transcend cultures and contexts and that, arguably, similar forces are exerted on every people

3 According to Mignolo (2006), the racial structure with which the imperial colonial differences have been historically founded [...] is the major impediment today to thinking seriously of global citizenship.

and society despite differing histories, cultures and languages. Thus, engaging international learners without separating them from their local contexts, online intercultural exchange provides unique opportunities for transformation that makes it possible for the knowing in the widest sense to occur, the knowing which provides "an opening up," which Heidegger calls "revealing" (p. 319).

Teaching plays a key role in this process. Educators, who are both masters of their subject matter and interdisciplinary open intellectuals and who see themselves as life-long learners and agents of globalization, become role models for students. Viewing theory and practice as firmly connected and mutually enriching, Kumaravadivelu (2012) rightly points out the need for teachers to be strategic thinkers and exploratory researchers. Also, as Dooly (2010) perceptively observes, teacher's competence becomes increasingly related to her or his ability to integrate Web 2. tools and to create a community of learners that extends beyond traditional classroom. One may add that the willingness and the ability to foster cross-national and cross-cultural understanding and acceptance, including through personal example, are indispensable for the successful instruction in the international classroom.

Thus, technology as "enframing" is instrumental for the *poiēsis* or 'making' of learning in the environment of telecollaboration. Online intercultural exchange, driven by learners who see themselves as agents of globalization and life-long learners, can become a manifestation of the liberating power of collaboratively created knowledge in any setting.

Telecollaboration and Innovation in Online Pedagogy and Education Research

Telecollaboration is an increasingly prominent topic in the works of scholars and practitioners, seeking innovative ways of promoting internationalization in education (see, for example, recent works such as Dooly, 2008; Guth and Helm, 2010; O'Dowd, 2012). One of the aspects these researchers highlight is that online intercultural exchange allows students to deepen their understanding of themselves and their own culture through the development of critical cultural awareness. This becomes possible, as O'Dowd (2012) suggests, when learners have opportunities in their online interaction to engage in intense periods of negotiation of meaning in which they can discuss cultural 'rich points' and elicit meanings of cultural behavior from 'real' informants in the target culture. Learners are also led to become more aware of the relativity of their own cultural beliefs and values as they try to make them explicit for their partners. However, researchers emphasize that this is only the case when online exchange involves explicit comparison of two cultures and the expression of direct opinions and reactions to the postings offered by others (O'Dowd 2006). Such dialogue between partners contrasts with interaction that involves an unreflective exchange of information between partners.

It is not by chance that pioneering practice and studies in intercultural online exchange have been conducted principally in the context of foreign language education. This is probably due to the fact that the whole concept of "interculturality," as Alred, Fleming and Byram (2006) note, stems from foreign language education. Over the years, however, this concept has been enriched by a variety of disciplines, including art, literature, psychology, education studies, cultural studies, political science and ethnography. Although a lot of research on telecollaboration has been written in the context of foreign language learning, intercultural exchange is by no means limited to this area of instruction. In fact, in 2010, the America Council of Education award "Bringing the World into the Classroom" recognized the innovative use of technology to promote internationalization through online intercultural exchange in such diverse disciplines as engineering, international business as well as science-oriented track for the international baccalaureate. All of these teaching experiments successfully infused an intercultural dimension into the curriculum in order to promote international learning.

Numerous developments in practice and research of telecollaboration underscore the importance of continuing in-depth investigation of the social, cognitive and linguistic aspects of this increasingly prominent learning environment which exemplifies the challenges and rewards of intercultural communication in education. With this in mind, we contend that telecollaboration is uniquely suited to serving social justice through intercultural democratic experience in education. According to Alred, Fleming and Byram (2006), this experience takes place when "people of different social groups and cultures engage in democratic social and political activity, not avoiding values and judgements" (p. 233). In this context "political" is taken in a broad sense to mean activity that involves working with others to achieve agreed goals. This experience, at its best, opens an opportunity to acquire subject-matter knowledge along with intercultural competence at a reasonable cost and without the need to physically leave one's own country.

In addition to formal instruction, travel and study abroad programs are often viewed as highly desirable for advancing intercultural communicative competence. As statistics indicate, however, only 1% of North American students per year (Goodman 2009) take advantage of this opportunity. As Bauman (2008) perceptively observes, "multicultural age" is the concept of the elite, while "the majority stays fixed to their place of birth" (p. 96). It is a task of democratic education to support this majority in their acquisition of intercultural competency, arguably indispensable for their success as democratic citizens and employees in the era of globalization, with its demands for tolerance, cooperation, and continuing self-renewal. In a spirit of solidarity and in attempt to promote social justice, one can see an urgent need for creation of alternative educational opportunities for these learners. As Lemke and van Helden (2009) point out, "we need to learn how to support the effective integration of learning across radically different activities, sites, institutions, media, networks and communities" (p. 169).

Telecollaboration, which unravels the importance of negotiation and understanding, along with personal values, is one of these opportunities. In this educational context, the need to acknowledge cultural strengths and uniqueness of each participant goes hand in hand with the openness to the creation of the supranational community of learners. This community, built on the grounds of the shared experience allows for the creation of "the relational capital," (Jackson, 2010) which is of high value in the 21st century, when local and world cultures become increasingly intertwined.[4] The acquisition of "stocks of understanding" (Jackson, 2010) requires tolerance and solidarity between speakers of different languages, who use language differently, when deliberating on different issues of interest. Tellecollaboration offers an opportunity to gain from these stocks through the pursuit of the education without borders.

---

4 I use the term "relational capital" along the lines of Jackson as a result of "accumulation of collective stocks of understandings that arise from the relational understandings of and between the different voices, histories and memories." (2010: 250).

Overview of the Book

The international team of the authors of these chapters is composed of researchers and practitioners who are currently located in the Middle East (Lebanon), Europe (Croatia, France, Poland, and the UK), and in North America (Mexico and USA). They span a range of identities, experiences and disciplinary expertise. Some base their chapters on empirical work, others on theoretical, and many on both: all view telecollaboration as a promising learning environment for mastering intercultural communicative competence, global understanding and acceptance. These skills can enhance content studies, and they are indispensable for personal and professional success of the individual in 21st century and for responsible world citizenship.

This book explores the contribution of telecollaboration to democratic education, solidarity and social justice in the globalized world as well as the complexities and challenges that arise from attempts to align international collaborations and social justice. It argues that a focus on internationalization and social justice should be integral to debates about the future of higher education. Drawing on their own innovative teaching practices, the authors develop and extend these debates from a multidisciplinary perspective and global contexts in the book's three interrelated sections.

Part 1, "Telecollaboration as a Practice of Inclusion" combines both Western and non-Western perspectives in the exploration of effective pedagogical and institutional strategies aimed at engaging students from the Middle East with international themes, at educating teachers from Eastern Europe for successful work in collaborative setting, and at preparing teachers of special education in the international learning environment.

Part 2, "Web 2.0 and Social Change" focuses on the new technological affordances, that in response to social, cultural and pedagogical imperatives, can enhance cross-border quality education and learning opportunities around the globe.

Part 3, "Intercultural Encounters as Transformative Experiences" turns to learners' identities and the impact of telecollaboration and technology- assisted instruction on the development of self.

The conclusion to the book returns to some of the key themes of its three sections, outlines possible directions for future research, and suggests that cross-border education, which we view as a form of intercultural exchange where all involved are both resources and hosts at the same time, should become an increasingly important dimension of openness in higher education. This dimension calls for solidarity and innovation that would meet challenges associated with the internationalization of education and social justice, and would bring the educators to the forefront in the struggle for peace and democracy in a globalized world.

References

Amirault R. J. and Visser Y. L. (2010) *The Impact of E-Learning Programs on the Internationalization of the University*. Nova Science Publishers, Inc.: New York.

Archer, L. (2007) Diversity, equality and higher education: a critical reflection of the ab/uses of equality discourse within widening participation. *Teaching in Higher Education*. 12(5): 635–53.

Bakhtin, M. (1993) *Toward the Philosophy of the Act*. Trans. Vadim Liapunov. Austin: University of Texas Press.

Bauman, Z. (2008) *Does Ethics Have a Chance in a World of Consumers?* Cambridge: Harvard University Press.

Alred, G., Fleming, M. and Byram, M (eds.) (2006) *Education in intercultural citizenship: concepts and comparisons*. Buffalo, NY: Multilingual Matters.

Castells, M. (1996) *The Rise of the Network Society, The Information Age: Economy, Society and Culture,* Vol. I. Cambridge, MA; Oxford, UK: Blackwell.

—. Informationalism, Networks and the Network Society: a Theoretical Blueprint. Retrieved 17 June 2013 from <http://annenberg.usc.edu/Faculty/Communication/~/media/Faculty/Facpdfs/Informationalism%20pdf.ashx>.

Daloz, L. (2000) Transformative Learning for the Common Good. In: Mezirow J. and Associates. (eds.), *Learning as Transfor mation*, pp. 103–125. San Francisco: Jossey-Bass.

Dewey, J. (1997) *Experience and Education*. New York: Simon and Schuester.

Dooly, M. (2010) Teacher 2.0. In: Guth S. and Helm F. (eds.), *Telecollaboration 2.0. Language, Literacies and Intercultural Learning in the 21st Century*. pp. 277–305, Bern: Peter Lang.

Dooly, M. (ed.) (2008*) Telecollaborative Language Learning.* Bern: Peter Lang Publishing Group.

Friedman, J. (1999) The hybridization of roots and the abhorrence of the Bush. In: Featherstone M. and Lash S. (eds.), *Spaces of culture, city, nation, world* , pp. 230–257, London: Sage.Freire, P. (1998) *Pedagogy of Freedom. Ethics, Democracy and Civil Courage*. Trans. Clarke, P. New York: Rowman and Littlefield. Goodman, A. (2009) Language learning and Study abroad: The Path to Global citizenship. *The Modern Language Journal* 93: 608–609.

Heidegger, M. (1993) "The Question Concerning Technology." In: Krell D. F. (ed.), *Basic Writings*. pp. 307–343, London: Modern Thought.

Heidegger, M. (2001) "Origin of the Work of Art." In: *Poetry, Language, Thought*. Trans. Albert Hofstadter, pp. 15–87, New York: Perennial Classics.

—. *An Introduction to Metaphysics*. Tr. Ralph Manheim. New Haven: Yale University Press, 1987.

Jackson, S. (2010) Learning through social spaces: Migrant women and lifelong learning in post-colonial London. *International Journal of Lifelong Learning*, *29*(2), pp. 237–253.

Kant, I. *"Idea for a Universal History from a Cosmopolitan Point of View"* Trans. Beck. L.W. Retrieved 18 June 2013 from <http://www.marxists.org/reference/subject/ethics/kant/universal- history.htm>.

Knight, J. and Altbach, P.G. (2007) The Internationalization of Higher Education: Motivations and Realities. *Journal of Studies in International Education*. 11 (3/4), Fall/Winter: 290–305.

Kumaravadivelu, B. (2012) *Language Teacher Education for a Global Society*. Abingdon: Routledge.

Lemke, J. and van Helden, C. (2009) "New Learning Cultures: Identities, Media and Networks." In: Goodfellow, R. and Lamy, M.N. (eds.), *Learning Cultures in Online Education*, pp. 151–170. London and New York: Continuum.

Mignolo, W.D. (2006) Citizenship, knowledge and the limits of humanity. *American Literary History*, 18(2):312–33.

O'Dowd, R. (2012) Intercultural communicative competence through telecollaboration. In: Jackson, J. (ed.) *The Routledge Handbook of*

Language and Intercultural Communication, pp. 340–356, Abingdon: Routledge.

Papacharissi, Z. (2011) *A Networked Self. Identity, Community and Culture on Social Network Sites*. New York and London: Routledge.

Pelletier, C. (2004) New technologies, new identities. The university in the informational age. In: Bayne S. and L and R. (eds.) *Education in Cyberspace*, pp. 12–25, London and New York: Routledge Falmer.

Tcherepashenets, N with Snyder, L. (2011) "Transformations: Life long Learners in the Era of Globalization." Jackson, S. (ed.), *Lifelong Learning and Social Justice*, pp. 190–210, *England and Wales: NIACE.*

Shirky, C. (2010) *Cognitive Surplus. Creativity and Generosity in a Connected Age*. New York: The Penguin Press.

Sousa Santos de, B. (1999) "Towards a Multicultural Conception of Human Rights." In: Featherstone M. and Lash S. (eds.), *Spaces of Culture. City, Nation, World. Spaces.* pp. 214–230.

Stetsenko, A. (2008) From relational ontology to transformative activist stance on development and learning: expanding Vygotsky's (CHAT) project. *Cultural Studies of Science Education* 3: 471–491.

Vygotsky, L. S. (1997) The history of the development of higher mental functions. In: Rieber, R. (ed.), *The collected works of L.S. Vygotsky* (Vo. 4, pp. 1–278). New York: Plcnum.

Part I: Telecollaboration as a Practice of Inclusion

Hayat Al-Khatib

Globalizing Online: Binary Distinctions or Global Partaking?

Introduction

Various types of telecollaborative practices have emerged with the opportunities provided by globalization in the information and technology supported communication age. Along with these practices, many practitioners and theorists have underscored the possibilities and implicit promises of uniformity and leveling of opportunities as individuals across the globe have engaged in intercultural dialogue. There are numerous examples of online, distance and e-learning practices that have utilized telecollaboration in advancing better learning processes and improving teaching and learning prospects (see Block, 2010, Al-Khatib, 2011). At the educational frontier, telecollaboration has been incorporated to maximize the opportunities afforded through technology and enrich the learning experience (see Jonassen *et* al; 2003; Fitzpatrick and Davies 2003; Chapelle 2004; Lauriero and Bettencourt 2011). Higher education institutions have begun utilizing online communication forums and venues to facilitate telecollaboration for working groups, with the aim of providing better learning experiences, opportunities for cooperating in educational and social activities and exchanging information at a global scale (see Knight 2005; Wilson *et al* 2007).

The technological innovation characterizing the twenty-first century has brought with it promises of improved globalized learning as new access is afforded and new frontiers explored, in the pursuit of better learning experiences, telecollaboration and improved educational practices. In an effort to utilize digital technology in advancing collaboration and better learning processes, educational institutions and especially universities in the developing world have begun promoting the use of technology in teaching and learning, based on the fact that

ICT resources and tools, the internet, multimedia and related technologies offer a wider arena for globalizing online, in pursuit of collaborative learning and the wider participation in internationalized learning experiences.

The opportunities offered by technical innovation have provided aspirations to communities and educational institutions to engage in collaboration in internationalized learning and aim for equality and social justice through capacity building and economic transformation. The role of ICT in the telecollaboration process has been perceived as contributing to an innovative learning experience and global partaking that moves beyond providing the link to distant parties to communicate, and paves the way for a wider landscape of genuine interaction drawing on multicultural resources. Arnold and Ducate (2006) referred to internet based intercultural exchange in the online meetings of EFL teachers. Dooly (2008) underlined the need to move beyond "a simple hook-up" (p. 66) and encouraged genuine interaction facilitated by telecollaboration. Palloff and Prat (2005) observed the development of collaborative skills at a socio-cognitive level, in groups associated with online learning in specific contexts. Additional benefits were related to the active construction of knowledge in participants through personal enquiry (see Cope and Kalantzis 2000; Barab, King and Gray 2004; Goodfellow and Lea 2007; Goodfellow and Lamy 2009).

Kress (2003) and Lauriero and Bettencourt (2011) proposed that the advances in technology have brought new potentialities for learning, to the advantage of the learner. The ease and spread of learning communities facilitated by digital mediums have provided opportunities for groups to collaborate at social, affective, cognitive and organizational levels and have taken education to global landscapes unconstrained by spatial or temporal limitations.

The Use of Telecollaboration in Various Online Education Programs

Education through telecollaboration and globalized online exchange has been recognized as the new phase of learning in the third millennium

(UNESCO 2005). The spread of global telecollaboration in education has been extended to include theories of comprehensive participation, internationalization and social justice (Sandholtz, Ringstaff, and Dawyer 2002; Wilson *et al* 2007; Bax 2011). According to Van Harmelen, telecollaboration in education is an informal and innovative approach that facilitates (2006) collective learning in contexts where new forms of technology are used in communication. Globalized online learning has been perceived as a way of introducing equal education opportunities around the globe (Dudeney 2007).

In developed countries, explorations in extending the physical teaching and learning spheres through telecollaboration have been correlated with positive outcomes for the learner (Dryer and Nel, 2003). In developing countries, the promotion of the internationalization of the learning experience through telecollaboration and globalized partaking is leading to the belief in the opportunity to acquire better learning experiences and high quality education. Internationalized education is thriving as the primary facet of this proposition. Several education enterprises embrace online learning and promoted the ideologies associated with telecollaboration (UNESCO 2005), e.g.

- Phoenix University has become the largest private university in the US, with online and on-campus provisions and programs ranging from sandwich nine week courses without certification, to Bachelor, Masters and PhD provisions. Within the university's online distance education program, telecollaboration is used to provide mentoring support.

- The Netherlands Business School has opened a branch in Nigeria (Universitiet Nigenrole). Telecollaboration is the primary mode utilized in exchanging information, comments and discussions between teachers and students.

- Harvard is developing two branch initiatives in Cyprus and the United Arab Emirates (UAE). Telecollaboration is promoted as the necessary medium of exchange for the participating cohorts.

- The University of Westminster is the key partner to the Kingdom of Bahrain University. It is also involved in similar partnerships in Nigeria, Uzbekistan and Kazakhstan. Telecollaboration is used in planning, monitoring and evaluating academic work.

- The United Kingdom Open University provides courses, collaborates on new curriculum, validates programs and shares expertise, through a long list of partnerships in open and distant learning. The Arab Open University is its Middle East partner. Telecollaboration is practiced at all levels; administration, faculties and students, to ensure the dissemination of information, timely follow up and the completion of tasks.

Although the process of telecollaboration and globalization online holds the potential of equality of access, to information, collaboration and world views, its application often reflects disparity between participants at more than one frontier. Inevitably, in this internationalization of learning, new concerns have started to surface and replace earlier, more optimistic perceptions. Researchers have brought attention to concerns about the ability of education through internationalized telecollaborative learning experiences to serve higher goals of access, equality and social justice, which have been reinforced by aspects of the digital divide, perceptions on cultural hegemony and the superiority of western thought (Bourne 2003; Leith 2003; Block 2009).

Online Education: Comparison between Developing and Developed Countries

Education through open and distance venues has been the innovative practice in teaching and learning in the new millennium (Kimber 2003; Bhagwati 2005; Knight 2006). Within the parameters of online education, telecollaboration through open and e-provisions has been perceived to relate to concepts of equality and social justice by offering practices of inclusion and securing participation across the globe (Jonassen 2000; Kimber 2003). From this perspective, social justice is becoming increasingly enshrined in educational practice, guided by an overarching focus on western thought and promoting equal access to education for all.

The demand for international education through online distance education has been increasing at a steady rate. Two million students sought international education in 2000. The number is predicted to increase to 7.2 million international students in 2025 (Fadlallah 2011). Part of this demand has been met by student mobility to international venues; the other proportion is achieved through international education, including telecollaboration in open, distance and e-learning modes.

The developed world has made use of the opportunities afforded to education, through information and communication technology to produce and disseminate knowledge and material (Aldrich 2003; Barab King and Gray 2004; Wilson *et al* 2007) to geographically distributed learners. The production and provision of new information relies heavily on globalized online communication, and the new mediums that are now available in the digital era. However, the rooted ideologies of the developing world, nourished by a long colonial history, have not always permitted engagement in collaboration on equal footing with institutions in the developed world. In addition, the diversity of learning methods in the developing world, that have been expanded by technological innovation, exposes learners and educational institutions to unexpected difficulties and challenges (Salmon 2002; Sandholtz, Ringstaff, and Dawyer 2002; Richards 2003, 2004; Fadlallah 2011).

One of the main areas of difficulty is the basic provision facing users in poor countries. The digital divide has materialized as a reality that confirms disparity in provisions, abilities and roles, during the practice of globalizing online (Fadlallah 2011). On the one hand, the digital divide between the developing and the developed world has proved to be a major obstacle to equal telecollaborative practice. Advances in information and communication technologies are inevitably related to access to provisions and while these are abundant in first world countries they are scarce in developing countries (Knight 2005, 2006). On the other hand, problems of cost, bandwidth and speed have been identified as further obstacles for access to information and communication in the developing world. Thus, the linking of telecollaboration to an improved learning process, conditioned by technological enterprise is hampered in the developing world by the economic and financial conditions, which in turn constrain the dream of equality as the gap grows wider (Bax 2011; Committee of Vice Chancellors 2002).

A more critical evaluation of the application of telecollaboration, which involves probing deeper into the pedagogy associated with such integration, underlines concerns that ICT supported learning is not able to fulfill wider pedagogic aspirations of transforming the learning experience in the developing world to ensure active collaboration charted by the learner. At the recipient end, the developing world utilizes the access provided to the global learning arena to accumulate knowledge, in a more passive way, without really engaging in selecting or disputing the content. Online education in third world and developing countries is a means of acquiring and assimilating foreign pedagogies and viewpoints without necessarily contributing to active engagement and knowledge build up. Telecollaboration, applied in an equitative manner in which all learners are engaged in equal opportunities to communicate, could help ameliorate this passive learning process.

Different concerns have materialized in the developed world. Europe, USA and Japan started engaging in a global competition to explore new trends and opportunities in the virtual landscape that embraces globalization and telecollaboration (Barab and Duffy 2004; Block 2010). The flow of culture, ideas, values, knowledge, technology and economy in one direction, through assumed telecollaboration exchanges has resulted in concerns on the cultural hegemony of western thought (Richards 2004; Norton 2000; Pavlenko and Blackledge 2003). Being focused on westernized themes and experiences, the provided material may not bear much relevance to the context of the developing country. Culturally sensitive considerations have to be applied in relation to what can be selected in terms of themes and debate topics.

Laurillard (2002), Gee (2003) and Kimber (2003) have provided a skeptical evaluation of success in using ICT in education and pointed out that despite its wider aim at inclusion and innovation, the pedagogy associated with the practice remains constrained, in the developing world, by a passive transmission model that alienates the learner and does not bring about the desired engagement. Cuban (2001) documents that in some contexts and with old school practitioners, telecollaboration intimidated rather than encouraged educators. Telecollaboration in this respect did not yield the expected improved learning experience (Lankshear and Snyder 2000; Laurillard 2002). The complexities and challenges that arise from attempts to align international collaboration in telecollaboration with improved learning outcomes have to contend with all the aforementioned difficulties.

The present landscape offers opportunities and challenges, entwined with different perspectives, on the dimensions of telecollaboration and international education. The growing debate centres around the learning experience and learners' reactions in relation to the implicit connotations of cultural superiority and digital disparity between learners in developing and developed countries (Block 2009; Bourne 2003; Goodman, Johnson and Webb 1997).

Binary Distinctions or Global Partaking

The Observatory on Borderless Higher Education (2002–2004) identified innovative and new developments in higher education. Education through telecollaboration can have different effects in different regions and contexts. It can have both negative and positive consequences, depending on a variety of factors. In the decades following large implementation of technology enhanced learning, the reassessment of the practice distinguished between the effectiveness of telecollaboration in technology enhanced mediums, the transacted content material and the associated perceptions and ideologies of the practice. The implicit values associated with reliance on transnational western material and ideology, exchanged through telecollaboration, in addition to the division of roles and provisions, has provoked the questioning of the neutrality of the venture.

The stark gap between the provision providers and the service recipient, and the strict division in the range and role of stakeholders impacts the roles and contribution of the collaborators and pose a central challenge to telecollaboration. "Cross-border education" encompasses a wide range of modalities in a continuum, shaped up by different forms of open and distance learning and using a range of technologies including e-learning (UNESCO, 2005). The process of integrating international, intercultural and global dimensions into the core of education through telecollaboration, coupled with the rigidity in allocating roles and privileges of collaboration are bringing new concerns and impediments.

The current study reflects on the experience of senior students in the Arab Open University in Lebanon, in an attempt to develop and extend the debate through focusing on dialogic modes of telecollaboration emerging from online interaction and supported by virtual forums.

The Context of the Study

The Arab Open University is a non-profit private university founded in 2002 in six regions; Kuwait, Jordan, Lebanon, Saudi Arabia, Egypt and Bahrain. In 2009 the seventh branch of Oman was added.

The university is funded by the Arab Gulf Programme for Development (AGFUND). The Arab Gulf Programme for Development (AGFUND) was established in 1980, upon the initiative of HRH Prince Talal Bin Abdel Aziz and with the support of leaders of the Gulf Cooperation Council Countries (GCCC): United Arab Emirates, Kingdom of Bahrain, Kingdom of Saudi Arabia, State of Qatar, Sultanate of Oman and the State of Kuwait. The Gulf Cooperation Council Countries (GCCC) pledged to support and finance projects in the Arab world, with a focus on human development, high quality education, alleviation of poverty and improving living conditions for vulnerable groups. One of these projects was the Arab Open University (AOU).

Through partnership with the United Kingdom Open University (UKOU), the Arab Open University AOU was able to offer programmes and courses across the national borders of seven countries to meet the demand for higher education. The cooperation with UKOU covers the licensing of material, consultancies, accreditation and validation. The Arab Open University has the faculties of Language studies (English Language), Business studies, Education, Information Technology and Computing.

The mission statement of the Arab Open University explicitly identifies aims of improving access and outreach to marginalized communities and people and providing better learning experiences. Technology-mediated learning, telecollaboration and globalizing online resources provided alternative and virtual forums to facilitate collective learning, supplement face to face tutorials and deliver education themes (Al-Khatib 2009).

The university invested much in e-provision to create suitable technologic infrastructure that was able to support the mission statement. Supported tuition was offered to provide access to socially disadvantaged groups with lower incomes as well as to larger sections of the society to help them gain access to equal opportunity and quality education.

The participating students in the study were final year students. They were required to complete a research project that involved using electronic discussion forums, online weekly meetings and mentoring support.

Extensive technical support was provided for faculty and student to facilitate use of technology in education and open and e-learning resources. The group comprised of 36 participants and 3 mentoring tutors.

Through utilizing multimodal methods of education, the Arab Open University was able to offer flexibility and provide access to sections of the society residing in remote geographic areas. The quest for social justice included reducing social discrimination based on income and elite education provision. High quality British programmes were validated and assured by reputable academic bodies (OUVS) and were perceived to increase opportunities of employability. E-provision included globalizing online, accessing support material, tutorial outlines, virtual conferencing, as well as online upload and download portals for assignment work. Telecollaboration and e-provisions reduced face to face attendance requirements and daily commuting from rural areas and encouraging interactive group work.

Table 1: Statistics: Gender ratio and graduate numbers in each branch.

Branch	*Females*	*Males*	*Total*	*Percentage of females*	*Graduates*
Kuwait	2989	2297	5286	56.55%	2691
Jordan	1447	953	2400	60.29%	2719
Lebanon	1745	2291	4036	43.24%	1175
Bahrain	714	740	1454	49.11%	355
Egypt	683	780	1463	46.68%	296
KSA	6530	6721	13251	49.28%	1647
Oman	821	781	1602	51.25%	—
Total	14929	14563	29492	50.91%	8883

Telecollaboration, Internationalization and Social Justice: An Analysis of the Program

The provision of IT support and technical infrastructure provided access to the wider web as well as virtual resources at no extra cost. The hybrid system of education at the Arab University offered access opportunity to applicants from rural areas, mature candidates and women. In particular, the double degree program on offer allowed an advantage in the world of work over other graduates with a single degree. The movement of knowledge, program, providers and curriculum across national and regional jurisdictional borders, facilitated by telecollaboration and the adoption of open, distance and e-learning, were related to goals of improving the learning experience and the higher ideological aims of inclusion and social justice.

The program was designed as an answer to the uni-directional flow of information, material and guidance through telecollaboration from providers in western universities coupled by requests for strict adherence to the prescribed program to fulfill degree requirement, resulting in some regions implicit restlessness and unease concerning perceived cultural hegemony. Furthermore, assumptions about the superiority of the material, propagated in some contexts by inexperienced facilitators, contributed to negative attitudes that associated the system with dichotomies of superior and inferior cultures. The introduction and practice of telecollaboration in open e-provision led to developing perceptions of equality and social justice among the participating cohorts. Learners and tutors appreciated the opportunities afforded by telecollaboration in and open e-learning environment that fomented independent learning. Within the wider program, telecollaboration provided a venue for debating local and international themes as well as an opportunity to assess the practice itself in an increasingly complex context. The discussion forum provided an opportunity to exchange views, discuss key issues and gain peer support in virtual learning contexts. Free access was provided as well as technical support through computer laboratories at the university cam pus. An activity report was extracted every week to monitor the involvement and progress of the group.

Most learners made use of telecollaborative tools to take part in the online weekly meetings, to exchange views and to get project mentoring support. Participants were organized into four forum groups in the shared virtual environments, based on their selected project macro-themes. The facilitator tracked the learners' progress, answered their questions, recommended useful resources and commented on their output. With time, learners assumed facilitators' roles in leading group discussion, initiating and collaborating in learning through the virtual learning environments.

Twenty-three members out of 39 participants took part in the discussion forum on a weekly basis. Eight participants were active on a regular basis while six participants never joined in the discussion forum.

The collaborative method offered participating students the opportunity to take the lead in their learning. Participants chose the project titles that reflected either immediate interests or future job orientation. They were encouraged to actively reform the generic modules and themes into titles that they perceived more relevant to their concerns. In other words, telecollaboration helped "liberate" the learners from tutor-centred instruction and gave them more space to negotiate their goals and plan their study route. They were more involved in the learning process and became partners in initiating discussions and exchanging comments on other group members' postings.

Forum	*Description*	*Discussions*	*Subscribed*
News forum	General news and announcements	2	Yes
Forum 1: Functional Linguistics Group (Systemic Functional Grammar)	May H. *Subtitling in 2012* (a study of authentic translation in films from English to Arabic) Rana D. *Functional Analysis of Rap* (Tu Pac *Changes* and Eminem *Foolish Pride*) Abir M. *Linguistic Features of E-mail Messages* Sahar Y. *How is the writer's persona preserved in translation?*	4	Top of Form 1 562 Yes Bottom of Form

Figure 1: Forum Access and Exchange.

Forum	*Description*	*Discussions*	*Subscribed*
Forum Two; Gender Theory (Sociolinguistics)	Elham M. *Gender Differences in Political Speeches (Luther King and Thatcher)* Therese A. F. *Comparative Analysis of Gendered Language in Men's Fitness magazine and Fit 'n' Style (ladies' magazine)*	2	Top of Form 1 563 Yes Bottom of Form
Forum Three: Language and Power (Critical Discourse Analysis)	Loubna K. *Power in Discourse and Power Behind Discourse (Bill Clinton Court Case*) Grace A. Z. *Racial Discourse in Lebanon* Fahda M. *CDA of Arabic and English news texts on CNN websites (reporting on Libya and Egypt uprising)* Nidal K. *The Role of Advertisement in Consumerism*	6	Top of Form 1 564 Yes Bottom of Form
Forum Four: Language Acquisition	Youssef S *Bilingual Acquisition* (*in 7–12 year old children*) Denise A. A. *The influence of the French language on writing in English (case study of 10 year old Lebanese trilinguals)* Sara R. *Language Analysis of English writing, case study of Grade 9 students.* Dalia H. *CMC and language issues*	3	

Figure 1: Forum Access and Exchange.

Forum One included students embarking on a career in translation. Forum Two was comprised of students wanting to pursue a career in journalism. Forum Three participants included a former lawyer and a human rights' activist. Forum Four attracted participation from elementary and intermediate cycle teachers.

It is notable that the selected titles grounded theory into contextual applications, e.g. racial discourse in Lebanon, critical discourse analysis of Arabic and English news texts on CNN reporting on the Libya and Egypt uprisings. These titles are related to increased cognitive and analytic abilities in the learners. Moreover, the postings added by the groups (Figure 2) conform to the active role assumed by the learners.

Discussion	Started by	Replies	Last post
Posting on Writer's Persona	Sahar G. 090776	2	Sahar G. 090776 Sun, 24 Apr 2011, 09:20 PM
Posting on subtitling in 2012	May H. 080793	3	May H. 080793 Fri, 22 Apr 2011, 01:42 PM
Posting on Functional Analysis of Rap (Tupac "Changes" and Eminem "Foolish Pride")	Rana F. 090023	2	Rana F. 090023 Fri, 22 Apr 2011, 11:04 AM

Figure 2: Participants and Postings.

Example: Extract A

Extract A reveals confidence in owning the knowledge propagated in the discussion as well as increased cultural awareness in relation to the subject under discussion.

> *Posting by Sahar G. 090776 – Sunday, 24 April 2011, 09:20 PM*
>
> Language users carry more than just the mere literal meaning in their discourse (either spoken or written); in fact, the choices they make on the syntagmatic and paradigmatic axes reveal several implicit meanings and details which are analised and detected through the Systemic Functional Liguistics of Halliday that studies language according to its functions.
>
> The research that I am conducting will try to study SFL in more applied contexts such as translation. I will try to answer the following questions:
>
> – Is the writer's persona preserved in translation and how?
>
> – Do the translator's lexical selections compromise the writer's persona in the target text?
>
> – Is the stance of the writer transmitted in the target text [stance refers to the attitude (positive, negative, neutral), modality (deontic, epistemic)]?
>
> – Is personalisation in the source text marked in the target language? How inclusive pronouns, pronouns of solidarity and direct addressing are reflected in the target text?
>
> – Is the standing of the author the same in the target text as in the source text?
>
> – Is the authenticity of the text related to the writer's persona?
>
> – Does the target text maintain the relationship established between the writer and the reader in the source text?
>
> -What are the effects of cultural norms on the writer's persona in the target text?
> II In order to find answers for the questions above, I will analyse the writer's persona of a text chosen from a book related to social studies and then translate the same text into Arabic (in three versions done by asserted translators) and study the text's translation to analyse the writer's persona in the target texts.

Example: Extract B

The student in Extract B discusses with confidence the racial problems that emerge through language use in a rap song.

> *Forum 1: Functional Linguistics Group (Systemic Functional Grammar) -> Functional Analysis of Rap (Tupac "Changes" and Eminem "Foolish Pride")*
> *by Rana F. 090023 – Friday, 22 April 2011, 11:04 AM*

My Project

Music has always touched hearts and minds. People have expressed the social ills of their particular era through popular music. Therefore, rap is a tool which allows the society to voice out thoughts about personal frustrations and sufferings. That is why, I decided to study how rap is not just some words said or introduced, but it is a worthy case in studying the language especially the lexicogrammatical features that rappers use in their songs to reflect specific situations or events they passed through. When rappers use language to express meaning, they do it in specific situation and for specific purpose "persuade, reject, etc..." Several questions are raised in this research: What structural features are placed in rap? Are they constant? Do they indicate any specific genre? What are the overall lexical and syntactical categorizations? Does it have any racial significant? And how?

According to Halliday, "language enables human beings to build a mental picture of reality to make sense of what goes on around them and inside them." And this is the case of rap songs.

According to Alicc Dcignan, mctaphors help people to talk about difficult, emotionally intense or uncommon experiences. Like the case of Eminem's rap song.

Deignan explains that speaker or writer may have chosen particular ways of expressing ideas metaphorically to convey an ideological or persuasive point. That is what the two rappers use to highlight their own experiences in order to deliberate a message or even to persuade others.

According to Fairclough, the conventions for speech acts which form part of a discourse type embody ideological representations of subjects and their social relationships, asymmetries of rights and obligations between subjects, these may be embedded in asymmetrical rights to ask questions, request action, complain, and asymmetrical obligations. Rap carries inside it speech acts through different representations like, request actions to end some case or raise other, complain on a specific issue (like racism, cheating,...), or ask for rights (like black people asking to be treated like whites)...etc.

The data that will be collected are written data of lyrics. The two lyrics belong to different rappers Eminem (the white rapper), and Tupac (the black rapper).

Findings and Conclusion

The extracts revealed learners as comfortable in discussing their themes and utilizing concepts from the taught linguistic theories. Participants made reference to racial and social issues, displaying their autonomy and competence in the subject of discussion. They took initiative, used examples from their learning context and expanded their area of analysis. Their discussions underlined their understanding of the taught material from the British program as well as their ability to manipulate wider applications.

Learners presented themselves as partners in the learning process; assessing, evaluating and critically engaging with the material. Moreover, the texts they selected did not contribute to views on cultural hegemony and the superiority of western thought but rather to an understanding of problem areas in other cultural contexts. The findings contribute to widening intercultural understanding which was afforded through telecollaboration and internationalizing the learning experience.

Using telecollaboration in open e-learning in this case study did not substitute indigenous knowledge by an alien one. On the contrary, telecollaborating in an open e-learning environment supported the development of an independent learning identity in participants, which in turn, contributed to the exchange of knowledge and information and reflected a growing interest and active participation in the learning process.

The telecollaborative experience brought improved learning experience and new-found confidence in the learners. Learners assumed active roles in the learning process and took initiatives in charting the path of their learning as well as in proposing topics and themes and responding to each other's' comments and feedback. This materialized through forum participation and telecollaboration in posting comments, exchanging views, discussing material, suggesting resources and in general, collaborating on project work through virtual forums.

Sources of potential learning have been expanded through telecollaboration, globalizing online and e-platforms. Learners assumed control and managed their learning. New ways of collaboration in learning, on the part of students, brought about new ways of monitoring and recognizing learning on the part of tutors. Learning moved from being a tutor-centered endeavor to becoming an authentic student-centered activity. Tutors changed their traditional instructors' roles and adopted facilitators' roles, with a focus on providing support for students in discovery-based learning. At group level, students assumed facilitators' roles in organizing and collaborating on the learning task.

Telecollaboration in this study underlined new approaches to assessing learning and globalizing online. Learners utilized telecollaboration to explore global themes in local contexts, e.g. racial problems were traced in Lebanon, as comparable to racial problems in the west. New found confidence in learners brought abilities to identify international themes and resulted in cross cultural understanding and improved learning experience.

The final reflections relate the success of telecollaboration in this case study to the ability of participants to move towards autonomy in learning, to participate in global themes and to build new roles. As for the participating tutors, the improved delivery modes that they experimented with, in addition to innovative guidance and flexible mentoring, provided enhanced learning experiences for the learners. If utilized well, telecollaboration in education can contribute to innovation, improved learning experience and intercultural understanding.

References

Arnold, N. and Ducate, L. (2006) Connecting Universities, constructing meaning: An analysis of beginning and future FL teachers "meeting" online. *Language Learning and Technology* 10(1): 42–66.

Aldrich, C. (2003) *Simulations and the Future of Learning: An Innovative (and perhaps revolutionary) Approach to E-Learning*. San Francisco, CA: Jossey-Bass.

Al-Khatib, H. (2011) Virtual realities, concrete results. *European Journal of Open and Distance Learning*. Retrieved 19 November 2013from<http://www.eurodl.org/materials/contrib/2011/Hayat_Al-Khatib.pdf>.

—. (2009) "How has pedagogy changed in a digital age?" the *European Journal of Open and Distance Learning* (EURODL).

Barab, S. and Duffy, T. (2004) From practice fields to communities of practice, in D. Jonassen and S. Land (eds.), *Theoretical Foundations of Learning Environments*, pp. 25–56, Mahwah, N.J: Laurence Erlbaum.

—, King, R. and Gray, J. (eds.) (2004) *Designing for Virtual Communities in the Service of Learning.* Cambridge, England: Cambridge University Press.

Bax, S. (2011) Normalization revisited: the effective use of technology in language education. *International Journal of Computer Assisted Language Learning and Teaching (I). CALLT* 1(2):1–15.

Bhagwati, N. (2005) *In Defense of Globalization.* New York: Oxford University Press.

Block, D. (2009) Researching language and identity in B. Paltridge A. Phakiti (eds). *Continuum Companion to Second Language Research Methods*, pp. 337–349, London. Continuum.

Block, D. (2010) Globalization and language teaching, in N. Coupland (ed), *Handbook of Language and Globalization*, pp. 287–304, Oxford. Blackwell.

Bourne, J. (2003) English for speakers of other languages, in Mercer and Swann (eds.) *Learning English,* pp. 287–304, Milton Keynes: Open University.

Chapelle, C. (2004) Technology and language learning: expanding methods and agendas. *System* 22 (1): 33–45.

Committee of Vice Chancellors (2002) *The Business of Borderless Education.* UK Perspectives. CVC P. London.

Cope, B., and Kalantzis, M., (eds.). (2000) *Multiliteracies: Literacy Learning and the Design of Social Futures.* New York: Routledge.

Cuban, L. (2001) *Oversold and Underused: Computers in the Classroom.* Cambridge, MA: Harvard University Press.

Disessa, A. (2000) *Changing Minds: Computers, Learning and Literacy.* Cambridge, Massachusettes: MIT Press.

Dooly, M. (2008) Understanding the Many Steps for Effective Collaborative Language Projects. *Language Learning Journal*, 92 (4): 614–620.

Dudeney, G. (2007) *The Internet and the Language Classroom.* Cambridge. Cambridge University Press.

Fadlallah, A. (2011) Bridging the gap in digital divide, in *The Fourteenth Cambridge International Conference on Open, Distance and E-learning*. Cambridge. Conference Proceedings.

Fitzpatrick, A., and Davies, G. (eds.) (2003) *The Impact of Information and Communication Technologies on the Teaching of Foreign Languages and the Role of Teachers of Foreign Languages.* EC Directorate General of Education and Culture.

Gee, J. (2003) *What Video Games have to Teach Us about Learning and Literacy.* New York: Palgrave Macmillan.

Goodfellow, R. and Lamy, M. (2009) *Learning Cultures in Online Education*. London. Continuum books.

Goodfellow, R. and Lea, M. (2007) *Challenging E-Learning in the University: A Literacies Perspective*. Maidenhead and New York. Open University.

Goodman, A. Johnson, P. and Webb, S. (1997) *Inequalities in the UK.* Oxford: Oxford University Press.

Jonassen, D., (2000) Revisiting activity theory as a framework for designing student-centred learning environments. In: D. Jonassen and S. Lund (eds.), *Theoretical Foundations of Learning Environments*, pp. 89–122, Mahwah, NJ: Lawrence Erlbaum.

—, Howland, J., Moore, J., and Marra, R. (2003) *Learning to Solve Problems with Technology: A Constructivist Perspective.* NJ: Pearson Education.

Kimber, K. (2003) *Technoliteracy, Teacher Agency and Design: Shaping a Digital Learning Culture*. Queensland University of Technology, Australia.

Knight, J. (2005) *Borderless, Offshore, Transnational and Crossborder Education.* October report of the Observatory on Borderless Higher Education. London.

Knight, J. (2006) Cross border education. In: J. Smart and B. Tierney (eds). *Higher Education Handbook of Theory and Practice*. Springer. Netherlands.

Kress, G. (2003) *Literacy in the New Media Age*. London: Routledge.

Lankshear, C. and Snyder, I. (2000) *Teachers and Technoliteracy*. Sydney: Allen and Unwin.

Lauriero, A. and Bettencourt, T. (2011) The extended classroom: meeting the students' needs using a virtual environment. *Procedia – Social and Behavioural Science Journal*. 15: 2667–2672.

Laurillard, D. (2002) *Rethinking University Teaching:A Conversational Framework for the Effective Use of Learning Technologies.* London: Routledge.

Leith, D. (2003) English: colonial to postcolonial. In: Graddol, D., Leith D and Swann, J. (eds.), *English History, Diversity and Change*, pp. 117–148. Dublin: Open University.

Norman, D. (2002) *The Design of Everyday Things*. New York: Doubleday.

Norton, B. (2000) *Identity and Language Learning: Gender, Ethnicity, and Educational Change*. Harlow, England. Longman. Pearson. Observatory on Borderless Higher Education (2002–2004) *Breaking News Stories*. OBHE London.

Palloff, R. M., and Pratt, K. (2005) *Collaborating online: Learning together in community*. San Francisco, CA: Jossey-Bass.

Pavlenko, A. and Blackledge, A. (eds.) (2003) *Negotiation of Identities in Multilingual Contexts*. Clevedon. Multilingual Matters.

Richards, C. (2000) Hypermedia, Internet communications, and the challenge of redefining literacy in the electronic age. *Language Learning and Technology*, 4(2): 55–77.

— (2003) *ICT – Supported Learning Environments: The Challenge of Reconciling Technology and Pedagogy*. Proceedings of international conference on computers in Education.

— (2004) From old to new learning: Global dilemmas. *Globalisation, Societies and Education*, 2(3): 399–414.

Salmon, G. (2002) *E-tivities: The Key to Active Online Learning*. London: Kogan Page.

Sandholtz, J., Ringstaff, C. and Dawyer, D. (2002) The evolution of instruction in technology- rich classrooms. In: R. Pea (ed.), *Technology and Learning*, pp. 255–276, San Franscisco: Jossey-Bass.

UNESCO (2005) *Guidelines for Quality Provision in Cross Border Education*. Paris.

Van Harmelen, M. (2006) Personal Learning Environments. In: Kinshuk, R. Koper, P. Kommers, P. Kirschner, D. Sampson & W. Didderen (eds.), *Proceedings of the 6th IEEE International Conference on Advanced Learning Technologies (ICALT'06)*. pp. 815–816, Washington, DC, USA: IEEE Computer Society Press.

Wilson, S. Liber, O. Johnson, M. Beauvoir, P. Sharples, P. and Milligan, C. (2007) Personal Learning Environments: Challenging the Dominant Design of Educational Systems. *Journal of e-Learning and Knowledge Society*. 16:1–2.

Anna Turula

Telecollaboration in the Clash of High- and Low-context Cultures: Social Justice at Stake

to Gosia Kurek, who helped me write this chapter in more than one way

Introduction

This chapter looks at selected collaborative and telecollaborative practices in Poland, focusing mainly on online co-operation in ELT teacher training. It concentrates on ways in which Polish learners of foreign languages, students of language studies and, in particular, future EFL teachers, perceive the nature and implementation of mono- and intercultural online exchanges. All this is considered in relation to theoretical assumptions as well as general expectations regarding telecollaboration presented in literature.

The main line of reasoning presented in this chapter rests on the assumption that if telecollaboration involves cultures in contact, there will always be both similarities and differences resulting from higher or lower cultural compatibility. While the chapter does not intend to question cultural similarities or underestimate their importance in learning, its main focus is on differences and resulting clashes, stemming from a virtually inevitable lack of coherence between culture-specific perceptions of both the general idea of what constitutes collaboration (or in this instance, telecollaboration) as well as the understanding of its different aspects. These aspects include, among others, the meaning of *working together* and the quality of individual on-task contributions, both of which will be given a closer look in the present chapter. Incompatibility has been found to be largely resulting from the fact that for some cultures collaboration towards a common goal may require placing the focus on task completion, whereas for others, group dynamics are of greater importance. There may also be discrepancies as to what the representatives of different cultures understand to be their obligation towards the (tele)collaborating party, and how the balance between me-and-my-culture vs. the-other-and-their-culture is established and maintained. This balance, whose nature is clarified in Section 1 as an equilibrium between appreciation of own culture and respect for the other culture, as well as equality in education that springs from such balance will be understood and referred to as social justice in the present chapter.

The approach to and implementation of every telecollaborative exchange can be said to be highly individual, rather than cultural, experience. This is difficult to dispute, particularly in the post-modern world with its global culture and fluid social boundaries. Yet – as I argue throughout the chapter – a national-culture tinge exists and influences the manner in which a group approaches telecollaboration, a tinge that is particularly typical of homogeneous societies such as that in Poland. This homogeneity can be both an asset and a burden. In the latter case, however, it is absolutely crucial to raise the awareness of Polish students – and future teachers in particular – to their own cultural specificity compared to that of more heterogeneous cultures. This can be done by providing them with a flexible understanding of the very concept of telecollaboration (including sensitivity to cross-cultural differences) as well as preparing them for actual participation in intercultural exchanges online by means of a number of activities, a selection of which are proposed and discussed in this chapter. The particular importance attached to such pro-telecollaboration training in the case of future teachers stems from a belief that it is one of the ways of enabling not only them, but more importantly, their prospective students to participate in intercultural dialogue on equal terms.

Telecollaboration vis à vis Bakhtinian Architectonics of Being: Its Potential for Social Justice

If two (or more) cultures come into contact, a number of interactive patterns can be identified, the dynamics of which will fluctuate in time.

Reasoning by analogy to the role of ego boundaries in language learning (see Guiora et al. 1972; Ehrmann 1999), I propose that – at least in the initial stages of the culture-culture encounter – participants with thick ego boundaries will be inclined towards own-culture-centredness and its resulting territorialism, while individuals whose ego boundaries are thin may be prepared to immerse themselves in the other culture and adopt (some of) its behavior patterns as their own. These two attitudes are extremities on an interaction-pattern continuum, a working model of which I propose in Figure 1, with in-between options of: *me-(and-the-other)* characterized by the need to present one's culture to the other party, usually for culture comparison, with own culture being central and seen as more valuable; the *me-the-other-and-me* perspective, when culture contact is a means of reflection on one's own culture, which is both a point of departure and the endpoint on such a mental journey; and, finally, *the-other(-and-me)* viewpoint based on a fascination of the foreign culture which is seen as being superior and has a potential to overpower own heritage. As was mentioned earlier, these different attitudes may be dynamic and sensitive to external influence to an extent determined, among others, by individual differences such as ego boundaries. These boundaries, even though person-specific, may be related to the type of culture the individual comes from. (This concept will be addressed later in this chapter)

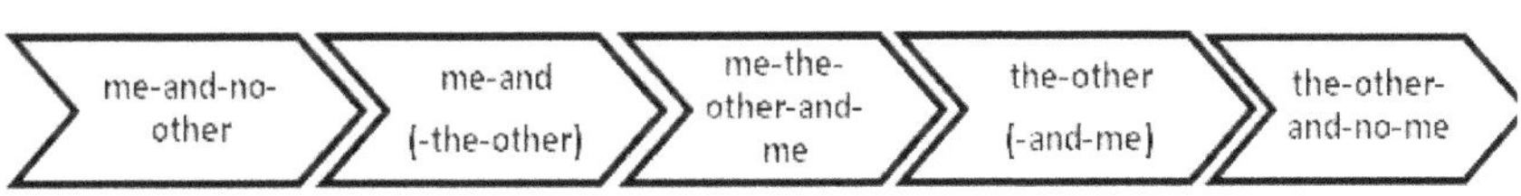

Figure 1: Cultures in contact: the interaction pattern continuum.

According to O'Dowd (2011), telecollaboration is "the application of online communication tools to bring together classes of language learners in geographically distant locations to develop their foreign language skills and intercultural competence through collaborative tasks and project work" (p. 342), and is first and foremost aimed at "[the reflection] on [the students'] own culture or their stereotypical views of the target culture" (p. 344). If we confront the above culture-continuum reflections with this definition, we arrive at the central point of *me-the-other-and-me*. While at this stage of reasoning our arrival is purely cognitive or theoretical, in the practice of telecollaboration this very same arrival will be the aim of all actions undertaken in the course of such culture-culture encounters. The process will be dynamic because it requires both (all) parties of the telecollaborative project to step out of the initial order of things and to assimilate a new order.

In light of this delineated model, telecollaboration will be, to use Bakhtin's words, "a two-faced Janus. It looks in two opposite directions: it looks at the objective unity of a domain of culture and at the never-repeatable uniqueness of actually lived and experienced life." (1999: 2). Yet, the harmony of things and acts of thinking, is created because "our intentions, imaginations, ideas and phenomena intertwine" (Cheung et al. 2007: 348). This happens only during the assimilation of the new order during which our initially bracketed consciousness has a chance to open and incorporate the new order of things. This refers, among others, to the specificities of the culture of *the other*, as well as one's own culture suddenly seen in the new, reflected light.

How does this happen? Following from the above-mentioned concept in conjunction with O'Dowd's definition, we can theorize that the opening of consciousness has every chance to result from the fact that telecollaboration is what Bakhtin (1999: 5) calls "the unitary context of a *subiectum's* once-occurrent actual life"; or a plane, on which the different participants' beliefs are confronted in the course of telecollaborative action/tasks. This process is dynamic and marked with tensions resulting from the fact that in every act of telecollaboration, there will – at least initially – be a demarcation line between the *normative* and the *factual*. The former is understood to be what the participants see as valid, based on "the grounds of the possibility of all experience" (Liapunov 1999: 81 in Bakhtin 1999). For a mono-cultural group taking part in a telecollaborative project, their previous own-culture-related experience will in fact be what Liapunov calls "all experience". This experience, as mentioned earlier, is bracketed in cognition, while the *factual*, in the form of telecollaboration reality, steps in and enters into dialogue with this set of held beliefs. This will result in a clash between *what ought to be* with *what is*. However, before *the ought* becomes an imperative to modify *what is*, the telecollaborative *factual* may have to be the context in which to raise the participants' cultural consciousness by questioning "the valid". In other words, in the course of telecollaboration its participants may come – or have to be gradually made to – realize that what they may see as universally valid is only locally valid, and what appears to them as "all experience" is actually "all of their experience."

What is important to emphasize is that the essence of the above-described change is in the opening of cultural brackets, and NOT in renouncing one's own culture (the attitude of *the-other-and-no-me* placed at the far end of the interaction pattern continuum; see Figure 1). While it is true that a smooth assimilation of a new reality does require new mental architecture, it does not mean that the old structure – determined by own culture – needs to be totally abandoned for the new-culture-determined structure of thinking to take hold. What it needs is restructuring and expansion. This is fully endorsed by Bakhtin's philosophy of the act in which, as he states (Bakhtin 1999: 15), "I empathize *actively* into an individuality and, consequently, I do not lose myself completely, nor my unique place outside it, even for a moment". The implication behind this statement is that the new structure should accommodate the understanding of both cultures in a mental construct which is new but in a specific sense. It is not a sum of the old perception of own culture and the acquired insights into the other culture, but rather a hybrid, a blend of entirely new quality. This is because "[e]mpathizing actualizes something that did not exist either in the object of empathizing or in myself prior to the act of empathizing, and through this actualized something Being-as-event is enriched (that is, it does not remain equal to itself)." (Bakhtin 1999: 15).

All of this is possible in general – and in telecollaboration in particular – owing to what Bakhtin calls *participative thinking*: engaged, committed, involved, concerned, interested, and essentially social, as:

> the standpoint of participation and involvement is the standpoint that we naturally occupy as social beings committed to participant relationships and acting under the sense of freedom, and it constitutes an understanding of objects or events that involves sharing or sympathizing with. (Liapunov 1999: 86)

Such thinking, which follows both the path delineated by O'Dowd (2011) in his definition of such practices and the route of Bakhtin's philosophy of the act, has every chance to result from telecollaboration. The participants of such online exchanges depart from their own culturally-determined routines (the given or Bakhtinian '*dannost'*) and embark on a journey through the cultural experience of otherness (given-as task; Bakhtinian '*zadannost'*) towards a new, restructured understanding of own practices in relation to the practices of another culture. In other words, and quoting Bakhtin again, they "seek to overcome [their] own givenness for the sake of what-is-to-be-attained" (1999: 11). This is what our answerable deeds or *postupky* within the Bakhtinian (1999) triada of the architectonics of being amount to. In this way social justice in telecollaborative encounters – and social justice in general – stands a chance of being achieved.

Culture Clash in Telecollaboration: Social Justice at Stake

As the previous section illustrated, if the human ability to empathize as well as the architectonics of being, based on the Bakhtinian triada of *I /the other / I-for-the-other,* are put to good use in telecollaboration, it has a potential for creating a plane for human encounters characterized by equality and social justice. Yet, I fully realize that this theoretical proposal is to a large extent idealistic, and the said potential may not always be entirely successful if confronted with (tele)collaboration reality. The main reason for this is the very same thing that constitutes the main value of telecollaboration: cultures in contact.

Following Hall (1976), we can distinguish two types of cultures: high- and low-context. The former are characterized by focusing upon in-group reliance and mutual support, a strong sense of tradition, history, marked language for humor expressions. The latter are oriented towards individuals: they value autonomy and personal freedom, so their social relationships may be looser or more superficial, characterized by considerable tolerance of ambiguity as well as openness to universal values (Hall 1976; Brown 1977).

The choice of underlying theory is by no means accidental. It is true that Hall's concept is essentialist, perceiving cultures as encapsulated and thus static, and, as such, has been subject to critique, especially when confronted with the post-modernist concept of culture dynamism, mobility, interaction as well as unpredictability and heterogeneity, typical of the present times often referred to as *liquid* (Bauman 2006, 2007). Yet, Hall's typology is applied here as a point of departure for two important reasons. First of all, while it is true (as I argue in the previous section) that culture/culture encounters are motivated by dynamically changing and individually determined attitudes of their participants as well as their individual differences (ego boundaries, for example), the national culture will definitely be part of such broadly understood context of every intercultural contact. The influence will be particularly consistent in homogeneous societies, of which Poland is an example. Obviously, it would be absurd to argue that Polish culture has not been subject to globalization. It has, yet it seems that the influence is mainly in the area of the humanistic rather than the social culture. For the latter, to become truly dynamic and heterogeneous, we, as a nation, would need genuine mobility (actual, not virtual) as well as more in-state cultural diversity. As Poland receives a fairly limited range of either of these culture de-capsulating experiences – or what can be metaphorically referred to as loosening of the atomic bonds in solids, whose rigidity is contrasted by Bauman (2007) with the fluidity of liquids – Hall's category of *high context* seems to be well suited as a basis for the description of our cultural specificity. On the other hand, the label of *low* context used here for the *liquid* social organization mainly for stylistic reasons, as a minimal-pair concept, seems to be justified too.

When we look at multi-cultural societies, with their constant dynamic interplay of all represented individualistic fragmented cultures, the liquidity is to a considerable extent compatible with Hall's definition of low-contextuality. Aspects of the latter, such as autonomy and personal freedom, go hand in hand with the post-modernist constructs of individual agency, whereas tolerance of multi-cultural societies and their openness to universal values result from the typically low-context lack of commonly recognized tradition and values as well as marked forms in different social rituals, one of which is language use. In other words, a truly heterogeneous multi-cultural society is very much low-context and, for the clarity of the argument presented, will be referred to as such in this chapter.

Related to context in culture is the concept of *face* understood, generally speaking, as our status or image. *Face* can be perceived literally (our actual face) as well as metaphorically (the impression we make); in online exchanges, where faces are rarely if ever truly disclosed, the *impression* or *demeanor* understanding of *face* will hold. It will also be motivated by the interpretation of the different definitions of *face*: Goffman (1959) argues that *face* are the masks put on and taken off depending on the audience, social context and interaction patterns; Lakoff and Johnson (1980) propose the *face for person* metaphor; Brown and Levison (1987) point out that *face* faces constant threats, which can be mitigated through politeness; for Lim and Bowers (1991) *face* is the public image we claim for ourselves (autonomy, fellowship, competence). What is important to the present line of reasoning is that *face* can be both personal and national (Brown 1977). In high-context cultures *face* is much more important and, consequently, more easily threatened as well as more fiercely defended. Face threats are posed, among others, by different positive and negative communication acts, such as (dis)approval (criticism or complimenting), directives (orders, requests), discussion of taboo topics, speech acts affecting interpersonal distance (confessions), etc. Face defense, in turn, will be manifested in a number of ways, some of which are discussed later in this section.

This poses a number of problems for telecollaboration. As a concept together with a considerable number of its practical implementations, it comes from the Anglo-Saxon culture, whose heterogeneity marks it as a low-context culture, and it is now being introduced to other cultural contexts, including high-context cultures, such as Poland. Bakhtin's claim that the participant of the act "intuits their inner lives as well as desires; he understands both the actual and the ought-to-be sense of the interrelationship between himself and these persons and objects" (1999: 30) may not fully apply to parties telecollaborating, as each of the cultures participating in an online exchange will have a set of *own* intuitions formed in the course of interpersonal, culture-specific encounters. These observations come from various pro-telecollaborative activities carried out in the course of ELT teacher training as a means of coaching for the intercultural online exchanges. The courses this relates to are: 1) *ELT Methodology* online, I taught in Spring 2012 to students of an MA English Studies program; and 2) *Collaboractive online*, a course in language as a social semiotic, taught to 11 volunteers, ongoing / former students of English Studies, which was still in progress at the time of writing.

The first type of problem to be observed in different aspects of telecollaboration are those arising in relation to the very nature of collaboration, especially some co-operation routines, which are seen as neutral in one culture, but may be perceived as face-threatening by another culture. This is likely to refer, among others, to feedback, which is an indispensable aspect of on-task work. The task-focused low-context participants may want to make it sincere, and, as such, it may be seen as hostile (and not constructive) by its high-context recipients. Conversely, comments provided by the high-context project participants are likely to be intentionally bleached of pragmatics and reduced to clichés so as not to be potentially threatening. For this reason corrective task-focused comments provided by foreign partners may trigger emotionally-loaded *face*-saving reactions in members of the Polish groups participating in telecollaborative exchanges. This hypothesis is based on the observation of collaboration between Polish students in local groups in which response to feedback happens to be that of denial and discreditation, as the students verbally express their unwillingness to consider their partners' suggestions, or even question their value (Sample 1).

Sample 1. Selected reactions to peer feedback to essay (tool: review; course: Collaboractive online)

Q: To what extent did you rely on peer feedback in the final draft of your essay?

A1: I didn't. I left the essay as it was, because I was more interested in the teacher's feedback.

A2: To very little extent.

A3: I didn't agree with the feedback.

A4: Most of the feedback seemed to be art for art's sake.

The answers quoted above show a very personal and emotional perspective of recipients on the feedback they had been offered and had seen as face-threatening rather than constructive and task-oriented. In turn, when offering critical comments, the representatives of a high-context culture often demonstrate a tendency to fall back on a routine which is referred to as social indifference in *face* literature (see Goffman 1967). It can be documented by examples of empty, safe feedback, such as the examples in the following sample of on-task discourse between students B and G, both of them participants of a collaborative project carried out as part of an online teacher-training course (Sample 2):

Sample 2: An extract from a collaboratively performed task: creating EFL age-related learning activities based on selected theoretical principles (task tool: email; course: ETL Methodology)

B (6/03/2012)

Hello!

I like Gosia's activity: simple, right for the pupils at that age.

I've just written 3 activities.

Look at them. I'm waiting for your opinions:)

G (7/03/2012)

Hi,

We are getting closer to the end of this project :)

I added one activity and attached the updated file. (Still one to go.)

It's not so easy to make them but I would be really interested to see later what other people included in their activities. I think it could be helpful for the future.

B (7/03/2012)

Hi!

I wrote my last activity.

What do you think about?

In addition to pragmatically bleached feedback, Sample 2 shows another cultural specificity: the understanding of the very concept of (tele)collaboration. The exchange quoted above demonstrates that the group's mode of task completion was not *collaborative* but what can be called *collective*: the assignment was carried out in a jigsaw-puzzle manner, the participants putting together their *individual* and *totally independent* contributions instead of working on the task in an *interactive* way. This collectivity in place of collaboration – which is the prevailing task execution mode in Polish groups – may result from yet another high- / low-context culture difference. As demonstrated by Rysiewicz (2011) in her study on how Anglo-Saxon and Polish EFL teachers conceptualize collaboration, the former see it as an instruction method, in which learners work in groups towards a common goal, while for the Polish educators collaboration is a team process where members support and rely on each other to achieve an agreed-upon objective.

The final example of the high/low-context influence can be the ways in which different participants perceive their own selves as seen in the format of online self-presentations. This is very clearly visible in a frequently implemented telecollaboration ice-breaking task in which course participants are requested to create group profiles either for the overseas partners or for their group mates. It has been observed – in my case based on the observation of profile updates in numerous e-courses I taught on the Moodle platform – that presentations created by the high-context Polish students tend to be very detailed personal accounts of social relationships. The face-enhancing strategies, previously mentioned in the context of evaluation, here usually take the form of excessive attention to the visual aspect of the presentations, in particular the participants' own images.

The observations made during the collaborative tasks carried out as part of the said courses allow for a comparison of the most conspicuous patterns of behavior as displayed by low- and high-context participants (Table 1).

Table 1: A comparison of high- and low-context patterns of behavior in telecollaboration.

High-context patterns of on-task behavior	Low context patterns of on-task behavior
emotional interpretation of tasks and activities	emotionally neutral perception of tasks and activities.
(over)reliance on social relationships	social relationships ancillary to task completion
important role of national and in-group face	focus on individual strengths and resulting contribution-to-task potential
subjectivisation of tasks/situations	objectivisation of tasks
territorialism; tendency to apply the *them vs. us* perspective	no / little territorialism.

What is important to point out in relation to Table 1 is that I see the comparison as tentative, a proposal of sorts, rather than a finite typology. This is because while the high-context patterns stem from observations I presented earlier, (see samples 1 and 2 as well as Rysiewicz 2011), the listed low-context regularities are not research-confirmed. They derive both from the earlier-described concept of *face* as well as the characteristics of TASK as it is understood in Anglo-Saxon (low-context-culture) ELT literature (see Candlin 1987; Nunan 1989; Long 1989) and are, potentially, a point of departure for prospective research.

The question that arises at this point is what kind of remedial action can be taken to improve the quality of telecollaboration between cultures of different contexts. This question is related to social justice in more ways than one. First of all, if we can successfully deal with the culture differences and resulting clashes, the telecollaborative encounters themselves will be more fruitful because, when not threatened, the participants are likely to be more willing to open the brackets of their consciousness. Secondly – and more importantly, especially in the case of TEFL training, which is of particular concern to me here – improving the quality of telecollaborative encounters is a condition *sine qua non* of preparing prospective teachers for undertaking and carrying out successful telecollaboration projects in their future professional careers. Only if trainee teachers learn how to smooth contacts between cultures of different context, can they, as teachers, be able to guarantee that their own students participate in intercultural dialogue on equal terms.

With regards to potential pedagogical solutions that may be applied in pro-telecollaboration teacher training, the pedagogic experience of a FL teacher suggests two instructional modes: inductive and deductive. Their implementation is presented in the following two sections. However, I would like to emphasize, with regard to both options, that the philosophy behind training prospective (language) teachers to smooth culture/culture contacts needs to be based on seeing cultural differences as a challenge to be benefited from rather than an obstacle to be overcome. As a result, whatever pedagogical action is undertaken, its expected result is the opening of cultural brackets, and *not* renouncing one's own specificity by adopting the norms of the partnering culture. This claim stems from a belief (Garfinkel 1967) that social integration and the stability of human encounters result not so much from uniformly accepted norms but from the ability to reflect on these norms and to seek an inter-subjective – rather than purely objective or subjective – perspective. As a result, the solutions described below focus on awareness-raising and opening the brackets of consciousness – and, potentially, loosening ego boundaries – rather than arguing for radical retraining and change of culturally-bound routines.

Training for Telecollaboration: an Inductive Approach

The most natural way of preparing students for socially equal participation in future intercultural dialogues is the experience of telecollaboration in its full, yet scaffold form, with the guidance and support of the tutoring teacher. For concrete examples of telecollaborative practices, readers may refer to a broad literature on the topic, including, among others:

- Tomé (2009), who describes an exchange between cultures in the form of weblogging;
- Hauck (2010), offering a description of an online telecollaborative project, with its special emphasis on *learning by doing*;
- Guth and Thomas (2010), who, in discussing different uses of tools in telecollaboration, mention such noteworthy telecollaborative practices as social bookmarking and intercultural media exchanges;
- Fratter and Helm (2010) showing how *savoir-être* – or the attitude of curiosity and openness to intercultural encounters – can be fostered in the course of collaboration and website creation;
- Dooly (2011), who provides an example of how language and content can be integrated in the course of an intercultural exchange online.

Each of the cited telecollaborative exchanges is an example of what is generally known as experiential learning. Understood as such, an actual telecollaboration practice will naturally go through all the four stages of Kolb's learning cycle (Kolb 1984). The concrete experience of such an intercultural exchange (stage1), especially if clashing with local routines, has a chance to be observed and reflected upon (stage 2). This is because the said clash is likely to open – or even force-open – the brackets of consciousness by pushing the participants out of their comfort zone into a territory whose novelty and initial incongruity with whatever concepts have been comfortably accommodated within one's consciousness, must result in some kind of reflection, which, as emphasized earlier, is most desirable in intercultural contacts. The next step is abstract conceptualization (stage 3) in which more general assumptions about the other culture (vis à vis own culture) are likely to be made. Ideally, these assumptions will result in the appreciation of heterogeneity rather than a tendency to modify own norms and dependent behavior. Finally, as a typical telecollaborative exchange lasts for at least a term, all the mental effort is likely to result in actions undertaken in the course of the following tasks, as a means of leaving the previously mentioned own-culture comfort zone and experimenting (stage 4) with new approaches.

This *do – reflect – conceptualize – do more – reflect again – … (ad finem)* cyclical mode is very much reinforced by task typology proposed by O'Dowd and Ware (2009), in whose sequence hands-on experience involved in self-presentation and product creation is interspersed with comparison, analysis and – again – reflection. An exchange based on the above-presented scheme (Kolb 1984) and taking the students step by step – not necessarily, but preferably – through 1) information exchange; 2) comparison and analysis; 3) product creation (O'Dowd and Ware 2009) serves as a good example of an inductive approach to intercultural training. As the experience of a number of such exchanges shows (see examples of telecollaboration quoted above; also Kurek, in press), a carefully structured sequence of telecollaborative tasks triggers highly authentic processes of negotiation, problem solving, evaluating and compromising, which in themselves are a valuable learning as well as culture-liquefying experience. What propels the learning process and broadens cultural horizons in the sense of facilitating attitudes of tolerance, openness to both *the other* and to the dynamics of social life in general, are the previously-mentioned culture-culture clashes and misunderstandings; they have been seen (Kurek, in press) to prompt genuine reflection well-accommodated within the Bakhtinian triada of *I* / *the other* / *I-for-the-other*. The participants of telecollaborative exchanges (re)discover their cultural identity and re-examine own beliefs and their limitations (*I*), by comparing them with those reflected in the approach to task presented by the partnering culture (*the other*); ideally – for the sake of smooth task execution as well as out of reasons that are more humanistic than purely pragmatic – while still preserving who they are, they learn to respect this different *other*, by making concessions as to how the task is executed and the interaction carried out (*I-for-the-other*). The nature of changes such as these is documented in Sample 3 (below), which cites the opinions of participants of the *Collaboractive online* who had had telecollaborative experience prior to enrolling into this course.

Sample 3: Samples of reflections on own telecollaborative experience (Task tool: email survey; course: Collaboractive online)

"Working in an international group is not easy – with all those different work styles and attitudes – but enriching in the sense that the differences make you more aware of your style and attitudes."

"Every contact with something new opens your mind and makes you more aware. This is what happened in the telecollaborative exchange in which I took part."

"You can see how other people approach the task and reconsider your own approach."

"I observed a difference in attitude (to class discipline) and reconsidered my respective attitudes. I'm not saying I changed my views – it just got me thinking."

"International collaboration online makes you sensitive to how others work – you need to respect this so that the task can be completed."

"I was impressed by how brave the other party was."

(About feedback)" It's educational to see oneself through the eyes of the other."

In this way, the process of intercultural collaboration enables its participants to expand their comfort zone and gradually broaden the brackets of their cultural consciousness to accommodate the new perspective within the existing conceptual model of society and its rituals (including language use) rather than abandoning their original norms and attitudes.

In relation to the above, it is important to note that although, in general, inductive approaches are traditionally considered to have a longer lasting learning impact than formal instruction, occasionally the tensions that intercultural encounters are fraught with may seem beyond resolution (Kurek, in press), and the tri-partite reflection model may be far from reality, especially in its third, *I-for-the-other* phase. In such cases, third-party moderation – in this case the teacher's – is needed. This may be of particular value to high-context learners, whose view of a given task and its performance may be emotionally tinged. As a result, as Helm and Guth (2010) point out, "mere information exchange can actually reinforce stereotypes rather than challenge them while the aim of telecollaboration should be to explore the multiple layers of each other's identities." (p. 26). Therefore, in such situations, teacher intervention is needed to shed new light on the participants' highly subjective feelings to help to put them in perspective in a broader social context. This is best done during in-class meetings, whose role in promoting reflection has been highlighted on numerous occasions (Müller-Hartmann 2000). This brings on the topic of the effectiveness of and the need for an alternative – deductive – instruction mode.

Training for Telecollaboration: a Deductive Approach

While the inductive approach is what best conforms with the Bakhtinian concept of participatory thinking, in multifaceted learning environments – as was mentioned in the previous section – the inductive training mode may not guarantee that all the different and important aspects will be addressed. This is particularly at issue in the light of the Noticing Hypothesis, based on which if something is to be learned, it needs to be noticed (Schmidt, 1990, 1991, 1995, 2001). Simultaneously, some individual predispositions (extrovertism; preference for analytical thinking; conformism – for a detailed specification see Turula 2011) require a more traditional, teacher-fronted type of instruction. I propose here that if such instruction is effective in TEFL (see Ellis 1994; Doughty and Williams 2004, to mention just the two main of the numerous works), it has a similar potential in (tele)collaborative tasks. An example of this efficiency is presented in a discourse snippet from collaboration on a task (Sample 4) performed *after* receiving teacher feedback on the previous unsuccessful attempts of this kind. This feedback included: (i) indication of the clichéd character of the remarks; (ii) examples of potential constructive on-task interaction and feedback (iii) suggestion of collaboration tools (Conceptboard; Google Drive).

Sample 4. Results of deductive instruction (reactive focus-on-form)

an extract from a student-student exchange during a collaboratively performed task (task tool: Conceptboard; course: ELT Methodology)

B: Probably it is a good example of an activity that refers to the isolation effect, but the first thing I came up with while reading the article was the task: circle the odd word out.

G: I think you are right. Your suggestion is better. The new task would be" Primary school children are given a few categories of words they already know. Each category includes an odd word out from the remaining words. Students are to cross out the word they think shouldn't be in a particular category." What do you think about it now?

B: Perfect.

***

B: Reading comprehension. The teacher suggests reading the questions before the text. Next students read the text on their own. When they finish the teacher asks one of the students to read the text aloud and then he/she asks comprehension questions to the students.

G: I don't know if it is a better version, but it's definitely the one which is commonly used during the lessons.

As Sample 4 illustrates, this on-task dialogue, when compared to the student interaction presented in Sample 2, is devoid of any clichéd character. Besides, there was proof of a synchronous task-related exchange (chat) showing that the task had been performed *collaboratively* (there was peer-to-peer work on task), and not *collectively* as in the previous attempt (the already-cited Sample 2).

In addition to such explicit instruction, which is called reactive, pro-active training is also an option. It may take the form of an on- / offline course, whose syllabus of 5 macro-tasks, each consisting of a number of component activities, is outlined below:

Deductive training for telecollaboration: a proposal.

1) Learning to provide constructive feedback
   (i) A collaboratively compiled list of what makes constructive feedback
   (ii) A collaboratively compiled list of empty feedback remarks (to be avoided)
   (iii) Practice (1): writing an essay assignment (topic related to TEFL) (optional: and indicating places in which you want other people to comment)
   (iv) Practice (2): commenting on another's work (avoiding emptiness in feedback)
   (v) A reflection on the feedback to Practice (1–2) – producing a collaboratively compiled list of the best comments: example – context – explanation (what makes feedback good)
2) Considering one's strengths and weaknesses: what I can offer; where I need help
   (i) Collaboratively compiling a checklist of predispositions needed in project work
   (ii) Self-reflection based on a checklist
   (iii) The Five Factor (OCEAN) Inventory of personality test[1] – as a reflection tool; correlation between the scores and checklist scores
   (iv) Collaborative analysis of the results of (2; iii)
3) Critical reflection on completed projects
   (i) Study of models; free-styling / offhand comments
   (ii) Delineating the structure: a collaboratively compiled list of what makes for good / bad collaboration on task
   (iii) A follow-up task designed by course participants
4) Reflecting on own culture in relation to other cultures in the areas investigated in 1–3
(namely: giving feedback, assessing one's contribution potential and actually contributing in project work); suggested form: mini-research into how a particular issue/ cultural aspect is seen through the eyes of other cultures. (in literature, films/ or newsrooms).
5) Actual telecollaboration

When it comes to the tools to be used in the implementation of the 5 macro-tasks, each of them should utilise different resources and social software (Dropbox, Google Drive, Conceptboard, ZOHO Project Manager, Edmodo, Facebook groups, WizIQ, etc.). This recommendation is made in the belief that offering "hands-on analysis of web resources and social networking tools" (Hauck et al. 2012: 82) leads to learner autonomy. Understood as "the informed use of a range of interacting resources in context (Palfreyman 2006 cited in Hauck et al. 2012), such independence and self-reliance is an important aspect of multi-literacy, part and parcel of raising cultural awareness, which, as pointed out repeatedly throughout the previous sections of this chapter, is the main aim of the undertaken pro-telecollaboration teacher training.

As for individual activities in the above-delineated course, several are provided below, to give the reader an idea of the character of such training. Each lecture is presented in relation to the relevant area of explicit awareness-raising while every task is discussed with reference to three levels of competence – organizational, pedagogic, technical – which prospective teachers need to be consciously sensitive to as well as needing to have a working knowledge of, if they are to become efficient facilitators of telecollaboration.

Deductive training for telecollaboration. Lecture 1

An online lecture on the pragmatic concept of face, face-threatening and face-saving acts, discussed in relation to high/low context in culture.

<u>Focus on</u> language and cultural awareness raising

<u>Tool used</u>: WizIQ

Deductive training for telecollaboration. Task 1

1) Students eavesdrop on people's conversations in the real world (and/or look at online exchanges at the forums) and put down samples of dialogue: human exchanges that involve risk (of being rejected / mocked / contradicted / etc.), filling in enclosed template.[1]

2) Students submit assignment via Dropbox.

3) Students examine ALL contributions, choose one category from the template and: (i) look for conversational patterns; (ii) reflect on how these patterns symbolize their own culture.

4) Teacher reads submitted reflections and offers socio-pragmatic feedback.

5) Students reflect on the task, evaluating its content and the content-tool matching.

Focus on competences:

competence: sensitizing students to the importance of keeping deadlines (in Part 3 everybody works individually, yet on material which is compiled collaboratively). The quality of individual work (*I*) as well as responsibility to *the other* is emphasized in the introduction to the task.

Pedagogical competence: raising language and cultural awareness; drawing attention to aspects of own language that reflect national culture stereotypes / social rituals, and can be a source of clash if transferred into the lingua franca used in a telecollaborative exchange and affect the *I-for-the-other* aspect of telecollaboration

Technical competence: learning to use and evaluating Dropbox as a content-sharing tool.

Deductive training for telecollaboration. Task 2

1) Students work in groups of 3–4 on lists of authentic teacher feedback shared with them via Google Drive. They are asked to look at the samples and: (i) categorize them into: constructive / empty; pleasant / unpleasant; (other binary – or not – categories of their invention); (ii) try to decide what makes the feedback constructive / empty; pleasant / unpleasant; etc; (iii) look at ways in which each comment could have been improved.

2) Based on Part 1 of the task, students write a 5–10-point list of guidelines on "How to give good feedback".

3) Additionally – and prior to task proper execution – they are supposed to decide (in the same Drive) how they are going to approach the task by collaboratively compiling (i) the code of practice; and (ii) a (provisional) timeline.

4) Teacher, who has access to both the process of the collaboration, as well as the end-product, provides feedback on both. If necessary, (s)he may choose to provide a model of good collaborative practice using Google Drive.

5) Students reflect on the task, evaluating its content and the content-tool matching.

Focus on competences:

Organizational competence: sensitizing students to the metacognitive level of task execution (planning; role assignment, etc.). An expected by-product is the reflection on both own strengths and weaknesses in collaborative work (*I*), strengths and weaknesses of other members of the group (*the other*) as well as the practice of remedial actions (if necessary) to be undertaken in task execution crises (*I-for-the-other*).

Pedagogical competence: raising awareness as to the importance of constructive feedback; analysis of what makes good feedback; immediate practical implementation of the knowledge in task execution, in all the on-task dialogic exchanges with other group members, combined with the reflection on how capable one is of giving and receiving such feedback (*I*); how others perform the task (*the other*) and how the language of feedback can be adjusted based on individual needs of its addressee (*I-for-the-other*).

Technical competence: learning to use and evaluating Google Drive as a collaborative writing tool with all the said dialogic / feedback options.

Deductive training for telecollaboration. Lecture 2

An online lecture on the different understandings of collaboration – collectivity vs. collaboration – in relation to the tradition of democracy in own culture vis à vis other cultures, with the foci on both, tradition of democracy and liquidity of culture. The emphasis on collaboration as *dialogue*, in which one needs to be brave (express own views; avoid pragmatic bleaching) and truly dialogic (*I-for-the-other* as a preferred attitude; deference before demeanor; creation of own image should always be second to task completion).

Focus on language and cultural awareness raising; focus on human relations
Tool used: WizIQ

Deductive training for telecollaboration. Task 3

1. Students do the FFM personality test. They learn their scores on the five OCEAN (openness, conscientiousness, extraversion, agreeableness, neuroticism) factors.

2. Based on a psychologist's feedback related to their test results, students keep a diary, in which they reflect: (i) on their strengths and weaknesses regarding collaboration; (ii) on their preferred roles in collaboration; (iii) on their preferences as to the proportions of individual and group work when on project.

3. The teacher comments on the self-reflection, mainly in the form of clarifying questions asked in relation to selected diary entries; the students are encouraged to comment back and enter into a prolonged dialogue with the teacher.

Organizational competence: sensitizing students to the meta-cognitive level of task execution (self-reflection on me-on-task; regularity the assignment incurs).

Pedagogical competence: directing the students' attention to individual differences in collaborative work and indicating the fact that every intercultural encounter will be dynamically motivated by personal as well as national factors.

Technical competence: the use of the *comment* function in text editors.

Going back to the overall learning model within which such tasks are accommodated, I would like to point out that the course described here is cumulative in the sense that its overall structure allows for the subsequent macro-tasks – (4) the culture reflection and (5) actual collaboration – to build on earlier activities 1–3. Moreover, the training has been devised in a way that makes it possible for the group to go from a predominantly teacher-centered, discrete, sequential set of activities to tasks that are more holistic and inductive in nature. What is also important is that while the first tasks (two of which have been presented in more detail above) concentrate, separately, on the project content, the language used in feedback as well as the different roles the participants may take on in collaborative work, the last two macro-tasks provide social context, in which the knowledge and abilities gained at the beginning of the course may be combined and used in a more gestalt way. As a result, the final activities demonstrate the quality of what Halliday (1978) calls "the semiotic structure of a situation" (p. 110) and are "a complex of three dimensions: the ongoing social activity, the role relationships involved and the symbolic rhetorical channel" (Halliday 1978: 110). In other words, the initial activities make the participants sensitive to Halliday's conceptual framework for representing this social context: field (project content; macro-task 3); tenor (project roles; macro-task 2) and mode (language used in collaboration; macro-task 1) for macro-tasks (4) and (5), in which the participants can "exchange meanings", and language functions as a social semiotic.

This final observation is of particular importance if we remember about the essential difference between the inductive educational experience of actual participation in a telecollaborative project (discussed in the previous section) and the deductive, teacher-fronted training scheme presented in this section. While the former will be authentic *per se* and, as such, is likely to ensure what Bakhtin calls "empathizing into an individual object of seeing – seeing it from inside in its own essence" (1999: 14), the latter can provide a similar learning plane within which the brackets of experience can be effectively opened only if the instructor takes care of the authenticity of the given tasks. This will include the following prerequisites: the presented problems should be genuine; the offered materials ought to have real-life quality; and the chosen tools need to be well matched to both the task and the participants' routines. Last but not least, the entire educational process must be highly individualized so that the learning experience is able to constitute a dynamic whole by being continuous and non-linear in a manner in which has every chance of being unique for each participant of such training. Only in such a case can "the performed act [will be taken] *not* as a fact contemplated from outside or thought of theoretically, but [...] from within, in its answerability" (Bakhtin 1999: 28).

As for the practicalities of implementation, the presented outline can function as independent pro-telecollaboration training; yet it may also be advisable to use its individual modules in the course of actual telecollaboration in cases of emergency should the teacher (or partner teachers) decide that their groups need scaffolding in some aspects of their collaborative work. This option also conforms with TEFL experience: it seems that the best result comes if both the inductive and deductive approaches are used alongside and in relation to each other. In a treatment model such as this one, the deductive element serves as a stepping stone (Rodrigo et al. 2002) for the participatory thinking in the course of actual task performance.

When it comes to the effectiveness of deductive training, it is difficult to generalize with the course – *Collaboractive online* – still in progress.

Yet, based on the data already gathered by means of a number of post-task feedback questionnaires, as well as self-reflection diaries written by the participants, it is possible to glean that the treatment has proven itself fairly promising in the area of opening the brackets of consciousness and sensitizing its participants to the dynamics of human interaction. Presented in Sample 5 are a number of answers to the regularly asked question *What have you learned*:

> Sample 5. Selected answers to the question 'What have you learned in the course of the activity?'
>
> (Task tool: email survey; course: Collaboractive online)
>
> "It was good to learn about 'the otherness' of fellow participants" (Introductory ice-breaking activity)
>
> "I learned to talk to strangers" (Introductory ice-breaking activity)
>
> "When you think about human interactions, you realize that it is all about the FACE. We need to preserve ours without damaging the face of the other" (Lecture 1)
>
> "Language is a powerful tool in human contact" (Lecture 2)
>
> "I better understood the role of dialogue with another human being as an eye-opening experience" (Lecture 2)
>
> "I learned collaboration modes are culture determined but it's not a sentence but a challenge (Lecture 2)"
>
> "Language symbolizes and reflects culture. That's why one needs to be careful how one uses it not to antagonize other cultures (Lecture 2)"
>
> "Doing tasks collaboratively makes our views, attitudes and working styles interweave (Task 2)"
>
> "I liked the constructive dialogue we engaged in when on task" (Task 2)
>
> "Respect for the values of your interlocutor and attentive listening to him / her are important aspect of collaboration" (Task 3)

Conclusions

Building on the idea of social justice understood in relation to Bakhtinian architectonics of being and the pragmatic concept of *face*, throughout this chapter I have argued that in order to be successful facilitators in (tele)collaboration, prospective teachers need prior preparation in this area either in the form of inductive hands-on experience or by means of deductive training. This is necessary due to considerable cultural differences, some of which result from high/low-context culture specificity. These differences, however, as emphasized in this chapter, pose a challenge rather than a threat to telecollaborative exchanges. A challenge, whose essence will be the liquefaction of cultural boundaries, especially in societies – and individuals – who are still rigid and atomistic in their approach to the *I – the other – I-for-the-other* architectonics of being. This liquefaction should amount to a greater tolerance, acceptance of the dynamism of culture and more openness in human relations. This, as it seems, is the very essence of telecollaboration.

As emphasized throughout this text, in the process of intercultural training and the said liquefaction of cultural boundaries, none of the cultures involved should be perceived as being the dominant one. Instead, the dialogical contact between culturally different mindsets that telecollaborative exchanges provoke, should lead to what was termed *me-the other-and me* standpoint, with the participant being able to reflect on their own culture without the feeling of *face* loss. This, as argued in the present chapter, is both the essence of telecollaboration and the foundation of social justice, the latter resting on the balance between *me* and *the other* as well as a flexible understanding of what *I-for-the-other* means for my own as well as for the telecollaborative partner-culture(s). In order to achieve and reinforce this understood social justice, we need language and culture awareness-raising in three areas. These areas are delineated based on Bakhtin (discussed thoroughly in the first section of the chapter) as: (i) our grasping a single fact as it happens; (ii) our understanding of its meaning; as well as (iii) the perception of ourselves in relation to it. In terms of pro-telecollaboration awareness raising, it means (i) conscious, reflective participation in online exchanges; (ii) an attempt to understand what individual telecollaborative activities mean to all involved parties / cultures; (iii) contemplation of different aspects of one's own culture in relation to other participants' cultures. Such consciousness-raising, as argued in the last two sections of the chapter, can result either from actual telecollaborative practice or be achieved in the course of a more deductive, teacher-fronted instruction consisting of a number of well-structured pro-telecollaboration tasks.

Finally, and very much in relation to the above reflection, I would like to emphasize that the two proposed solutions (see the last two sections of the chapter) to culture clashes (described in the second section), have been considered from my own national and cultural perspective. This, however, does not mean that I regard remedial actions and the resulting improvement of the quality of (tele)collaboration as the sole responsibility of high-context cultures, one of which I, myself, represent. In other words, it has never been my intention to argue that it is high-context cultures, which are new to (tele)collaboration, that need to adjust to low-context cultures, in which the very construct originated. Quite to the contrary. First of all, such a demand would contradict Bakhtin's claim that losing oneself and one's cultural specificity results in impoverishment rather than growth. Secondly, and more importantly, I believe that social justice and equal-term participation in cultural exchanges online can be guaranteed only if *both* cultures look critically at the architectonics of being through telecollaborative encounters (a question in need of further research into certain on-task routines, suggested in the second section). Critical analysis should amount to reflective consideration and not (self-) deprecation. This means reflecting on ways in which high-context cultures can potentially enrich the original experience of such online exchanges. They can do so by offering their cultural specificity, in my case, Polish values, such as solidarity as well as mutual reliance and support. In other words, while culture – culture contact is likely to liquefy the boundaries of homogeneous culture, it is also true that the atomic bonds of human relations typical of high-context societies – including the notion of friendship, whose semantic constitution differs between cultures (Wierzbicka 1997) – may be what offers a sense of stability and mutual reliance in what Bauman (2007) sees as the uncertainty of the times of liquefaction.

References

Bakhtin, M. M. (1999) *Toward the Philosophy of the Act. Translation and notes by Vadim Liapunov. 3rd edition.* Liapunov V. and Holquist M. (eds.), University of Texas Slavic Series No. 10.

Bauman, Z. (2006) *Liquid Modernity. 7th impression.* Cambridge: Polity Press.

Bauman, Z. (2007) *Liquid Times. Living in an Age of Uncertainty.* Cambridge: Polity Press.

Brown, B. (1977) Face Saving and Face Restoration in Negotiation. In: D. Druckman (ed.), *Negotiations: Social-Psychological Perspectives.* pp. 275–300. Sage.

Brown, P. and Levinson, S.C. (1987) *Politeness: Some universals in language usage.* Cambridge: Cambridge University Press.

Candlin, C. (1987) Towards task-based language learning. In C.Candlin and D. Murphy (eds): *Language Learning Tasks.* pp. 45–48. Englewood Cliffs, N.J.: Prentice Hall.

Cheung, Ch-F., Loparic, Z., Copoeru, I., Sepp, H.R., and Embree, L. (eds). (2007) *Phenomenology 2005, v.4,* Zera Books.

Dooly, M. (2011) Divergent perceptions of telecollaborative language learning tasks: task-as-workplan vs. Task-as-process. *Language Learning and Technology* 15 (2). 69–91.

Doughty, C. and J. Williams (eds.) (2004) *Focus on Form in Classroom Second Language Acquisition.* Cambridge: Cambridge University Press.

Ehrman, M. (1999) Ego boundaries and tolerance for ambiguity in second language learning. In J. Arnold (ed.) *Affect in Language Learning.* pp. 68–86. Cambridge: Cambridge University Press.

Ellis, R. (1994) *The Study of Second Language Acquisition.* Oxford: Oxford University Press.

Fratter, I. and Helm, F. (2010) The Intercultural Project. In: Guth, S. and Helm, F. (eds), *Telecollaboration 2.0. Language, Literacies and Intercultural Learning in the 21st Century.* pp. 385–398. Peter Lang: Bern.

Garfinkel, H. (1967) *Studies in Ethnomethodology.* New York: Prentice Hall.

Goffman, E. (1959) *The presentation of self in everyday life.* Garden City, NY: Doubleday.

Goffman, E. (1967) *Interaction ritual. New York: Anchor Books.*

Guiora, A., Brannon, R. and Dull, C. (1972) Empathy and Second Language Learning' *Language Learning*, 22. 111–130.

Guth, S. and Thomas, M. (2010) Telecollaboration with Web 2.0 tools. In: Guth, S. and Helm, F. (eds), *Telecollaboration 2.0. Language, Literacies and Intercultural Learning in the 21st Century*. pp. 39–68. Peter Lang: Bern.

Hall, E.T . (1976) *Beyond Culture*. Anchor Books.

Halliday, M.A.K. (1978) *Language as a Social Semiotic*. London: Edward Arnold.

Hauck, M. (2010) The Enactment of Task Design in Telecollaboration 2.0. In: Thomas, M. and Reinders, H. *Task-Based Language Teaching and Technology*. pp. 197–217. London: Continuum.

Hauck, M., Fuchs, C. and Müller-Hartmann, A. (2012) Promoting learner autonomy through multiliteracy skills development in cross-institutional exchanges. *Language Learning & Technology 16, 3*. 82–102.

Kolb D.A. (1984) *Experiential Learning experience as a source of learning and development*. New Jersey: Prentice Hall.

Kurek, M (in press) The potential of telecollaborative projects for teacher training programs.

Lakoff, G. and Johnson, M. (1980) *Metaphors We Live By*. University of Chicago Press.

Liapunov, V. (1999) Notes. In: M.M. Bakhtin, *Toward the Philosophy of the Act. Translation and notes by Vadim Liapunov. 3rd edition.*

Lim, T.S. and Bowers, J.W. (1991) Facework: Solidarity, Approbation, and Tact. *Human Communication Research* 17, 415–450.

Long, M. (1989) Task, group and task-group interaction. *University of Hawaii Working Papers in English as a Second Language 8*. 1–26.

Müller-Hartmann, A. (2000) The role of tasks in promoting intercultural learning in electronic learning networks. *Language Learning & Technology, 4*(2):129–147.

Nunan, D. (1989) Understanding Language Classrooms. Englewood Cliffs, N.J.: Prentice Hall.

O'Dowd, R. (2011) Intercultural communicative competence through telecollaboration. In: J. Jackson [ed.] *The Routledge Handbook of Language and Intercultural Communication*, pp. 342–258. London, Routledge.

O'Dowd, R., & Ware, P. (2009) Critical issues in telecollaborative task design. *Computer Assisted Language Learning*, 22(2), 173–188.

Palfreyman, D. (2006) Social context and resources for language learning. *System* 34(3), 352–370.

Rodrigo, M.J., B. Triana and M.I. Simón. (2002) Cognitive variability in the development of the concept of family: a contextualist or the gradualist view. In: M. Limón and L. Mason [eds], *Reconsidering Conceptual Change. Issues in Theory and Practice*. pp. 165–186. The Netherlands: Kluwer Academic Publishers.

Rysiewicz, J. (2011) *The Influence of Individualism and Collelctivism on the Teachers' and Students' Perspectives Regarding Collaborative Learning – the Case of Poland and Foreign Countries*. Unpublished MA dissertation.

Schmidt, R. (1990) The role of consciousness in second language learning. *Applied Linguistics,* 11: 129–158.

Schmidt, R. (1991) Input, interaction, attention, and awareness: The case for consciousness-raising in second language teaching. *Anais do X Encontro de Professores Universitários de Língua Inglesa, 1.2.* pp. 1–23. Rio de Janeiro: PUC/RJ.

Schmidt, R. (1995) Consciousness and foreign language learning.: A tutorial on the role of attention and awareness in learning. In R. Schmidt (ed.) *Attention and awareness in foreign language learning*. pp. 1–64. Honolulu, HI: University of Hawaii, Second Language Teaching and Curriculum Center.

Schmidt, R. (2001) Attention. In: P. Robinson (ed.), *Cognition and Second Language Instruction*. pp. 3–32. Cambridge: Cambridge University Press.

Tomé, M. (2009) Productions orales, *weblogs* et projet de télécollaboration avec le *web 2.0* pour l'enseignement du français (FLE), *Revue ALSIC*. 12 : 90–108.

Turula, A. (2011) *Form-Focused and the Advanced Language Learner. On the Importance of the semantics of grammar.* Frankfurt: Peter Lang.

Wierzbicka, A. (1997) *Understanding Cultures through their Key Words: English, Russian, Polish, German, and Japanese*. New York: Oxford University Press.

Rebekah McPherson, Kelley King, Andrew Milson,
and Mary Bailey Estes

Evaluation of an International Online Learning Initiative in Special Education

In spite of concerted efforts through traditional means to increase the supply of special educators, shortages persist both in the US and abroad (Spooner *et al* 2000); therefore, teacher preparation programs are turning to online education to provide equitable access (Kurtts and Vallecorsa 1999; Mohr 2004). The increase in distance learning (Ludlow 2001) is revolutionizing training for special educators, and its use could "systematically impact the shortage" (Spooner *et al.* 2000: 92). Literature suggests that distance learning in special education teacher preparation has the potential not only to reduce the shortage (Spooner, Algozzine and Jordan 1998; Smith and Meyen 2003; Johnson 2004), but also to provide the most up-to-date training, especially in rural and remote areas (Ludlow *et al* 2002; Johnson 2004).

Scholars believe that distance education for special educators provides a new form of pedagogy (Meyen *et al* 2002; Smith and Meyen, 2003), offers "anytime, anywhere" access to professional development (Meyen 2003: 84), increases special education certification enrollment (Fore, Martin and Bender 2002), reaches a global audience (Sun, Bender and Fore 2003), increases the enrollment of non-traditional students (O'Neal *et al.* 2007), and reduces stress and burn out (Ludlow *et al.* 2002; Sun *et al.* 2003). Other advantages of distance education include its allowance of effective strategies for field experiences (i.e. cybervision) (Binner and Falconer 2002; Lignugaris and Kraft 2002; Jung *et al.* 2006; Knapczyk and Hew 2007) provision of opportunities where none or few exist (Spooner *et al.* 1999), building collaboration among educators (Bore 2008), and online mentoring capabilities (Knapczyk *et al.* 2005).

Findings from several studies, specifically related to online education for special educators, served to support our research. Knapczyk et al. (2005) reported positive results from practicum teachers and mentors using online mentoring. Bore (2008) surveyed students after their online course and found that they viewed taking another online class favorably. Huai *et al.* (2006) failed to confirm that the level of prior Internet competencies impact gains in online learning. Steckelberg *et al.* (2007) developed one online model and successfully shared the resources with six other university sites, suggesting that outside programs can be implemented effectively in other institutions. Sharing Steckelberg *et al.*'s views, Smith and Meyen (2003) explained how some online teacher preparation programs are teaming up to share courses or Reusable Learning Objects (RLO). RLO implementation provides a much greater volume of resources than an individual faculty member could develop alone. Jameson and McDonnell (2007) encourage universities to have students complete programs as a cohort and provide an initial orientation training to help students feel comfortable with the online system. Menlove and Lignugaris/Kraft (2004) also suggest that orientation training in WebCT may lead to lower attrition. Knapczyk *et al.* (2001) Collaborative Teacher Education Program produced a co-operation and collaboration among students that lasted past the program period. Rowlison (2006) reported that students' most frequent comment was that they liked being able to work at their own pace.

Distance learning programs for special educators worldwide is fairly new. Ludlow and Brannan (1999) reviewed programs from 1985 to 1999 and found no references to international program offerings. The first global initiative was Project STARS in 2002 at West Virginia University (Ludlow and Duff 2002). In the same program in 2009, Ludlow and Duff reported enrolling 10 students from around the nation, as well as six individuals living abroad from Japan, Guam, Saipan, Iceland and Costa Rica (Ludlow and Duff 2009). The STARS program is only offered in English (Ludlow and Duff 2009). In another program, Project ACCESS, two professors on the Greek island of Cyprus delivered courses to students in New Mexico and Colorado (Kelly and Schorger 2003). Finally, there are the courses that are the focus of this research. The University of North Texas and Universidad Casa Grande in Ecuador formed a collaborative partnership to allow full-time teachers in both countries to take online courses together. Coursework was offered in English and Spanish (Tyler-Wood, Barrio and Peak 2009). The Internet's impact on the internationalization of higher education continues to shape course offerings. Developing strategies for building collaboration among a diverse group of international learners and culturally responsive curriculum is essential as more Western universities package curriculum that is shipped world-wide via the Internet (Rossman, 1992, Cerny and Heines 2001; Onay 2002;) and institutions extend enrollment to students abroad (Gunawardena, Wilson and Nolla 2003; Rasmussen, Coleman and Ferguson 2006; Rogers, Graham and Mayes 2007).

This project addressed the social justice issues of improving the quality of international educational resources for special education teachers during a time of severe global shortage and the compelling need to train educators how to teach increasingly diverse student populations (UNESCO, 2006). The Center on Personnel Studies in Special Education reported that in the 2000–01 school year around 98 percent of school districts in the United States reported shortages of qualified special education teachers, noting that "approximately 47,500 special education positions were filled by uncertified personnel — a 23 percent increase from the previous year" (COPSSE 2004: 1). Special education teachers from under-represented ethnic groups are on the decline; consequently, the current pipeline of White, middle-class, monolingual female students must be trained in diversity and multicultural best practices (Zeichner 1993; Banks 2001; Sleeter 2001; Yzquierdo, Lopez-Reyna, and Flippin 2002; Ladson-Billings 2005). In the United States, the continuing growth of the Spanish-speaking population, as well as the disproportionate placement of minority students in special education (Coutinho and Oswald 2000) heightens the need for educators who understand the needs of bilingual students in special education (Cummins 1989).

The economic downturn of 2008–2012 worsened the educational situation in developing countries. In Ecuador, UNESCO's Regional Overview for Latin America and the Caribbean reported that only 71% of teachers were pedagogically trained (2009), this is a 16% decrease from data gathered in 2000 (UNESCO 2006). Even in hard times, Ecuador is making tremendous strides in providing special education services to students. To facilitate these advances, educators request not a handout, but the opportunity for more training (Stuecher and Suarez 2000). Both in the US and abroad, the shortage of special educators and the need for educators who are trained in teaching diverse student populations might be curbed by the increased availability of specialized online resources.

Critical unanswered questions in the literature are: How will the Internet be employed in special educator preparation, not only in the US, but abroad (Ludlow and Duff 2009)? How can multicultural awareness attitudes be developed in future and current educators (Nieto

2009)? How can distance-learning programs for special educators be evaluated to promote continuous improvement (Bore 2008)? In addition, information was lacking on the use of telecollaboration to prepare more globally aware special educators, research on special educators in Ecuador and the current climate in special education, and concerted efforts of US institutions to provide training to meet the global special educator shortages. In a limited way, our work addresses these needs. Researchers have conducted program evaluations of distance learning programs in the US; still, very little research exists on the evaluation of international online special education programs. Moreover, no studies could be found involving bilingual, online special education courses. Although one study does not provide all the answers, these findings contribute to the literature by evaluating two international, bilingual, online special education courses, when previous research was limited or nonexistent.

This study consisted of a formative program evaluation of the first two courses of a four-course sequence within a bilingual web-based special education master's program. The research was conducted at the halfway point of the four-course sequence. The University of North Texas offered the online courses to students in the United States and to students abroad in a private college located near Ecuador's western coast in Guayaquil, Ecuador. The courses, which were still in development, allowed special educators from both countries to engage in learning and participate in telecollaboration because all the course materials were provided in English and Spanish within the same Blackboard interface. A global initiatives grant from the US university funded the four courses. The aim of this research was to conduct a formative program evaluation to identify improvements for the beginning courses, while the courses were still capable of being modified, in the hopes of improving the quality of the courses prior to offering the courses to a wider audience. The strengths and weaknesses of the implementation of one of the course goals related to enhancing teachers' multicultural awareness were examined. This goal was approached from two perspectives, the first being how the American institution integrated the multicultural course content; and the second, how the US institution employed cultural sensitivity in adapting the courses for the Ecuadorian students.

This formative program evaluation was guided by the following research questions:

1) What is the evidence of multicultural awareness in these two special education courses?
2) What is the cost-benefit analysis of the course for the institutions?
3) How was the underlying course goal of enhancing students' multicultural awareness implemented in the course?
4) What are the unintended positive or negative consequences of the course design and implementation?

Materials and Methods

Participants

Description of the participants from the general-orientation course, Special Education Programs and Practices. Sixty students were enrolled in the course, 34 from the US and 26 from Ecuador. Of those 60 students, 20 (39%) responded to the online learning survey. Nine participants were from the US institution and 11 were from the Ecuadorian university. All participants were female. The majority of students were between the agcs of 26–30, had two years or less of teaching experience, attended classes part-time, and responded that they were familiar with online learning.

Description of the participants from the specialized course, Assistive Technology. Thirty-six students were enrolled in the course, 11 from the US and 25 from Ecuador. Of those 36 students, 15 (41%) responded to the online learning survey. Four participants were from the US institution and 11 from the Ecuadorian university. All participants were female. The majority of students were between the ages of 26–30, had 3–5 years of teaching experience, attended classes part-time, and responded that they were familiar with online learning.

Description of the semi-structured interview respondents. The selection of interviewees was based on respondent availability. Six interviews were conducted. To capture the voice of the US institution, the four-course sequence course developer, the Assistive Technology course instructor, and the bilingual teaching assistants for both courses were interviewed. To satisfy the perspective of the Ecuadorian university, the Dean of the

College of Education and the program coordinator for the collaborative initiative were interviewed.

Course Information

The official course description explains that EDSP 5710 was an introductory course. In order to enroll in the general-orientation course, UNT students completed the general admission requirements. There were no prerequisites for 5710. Several sections of the course were usually offered during the fall and spring semesters; however, UNT only offered one 10-week section during the summer of 2008. UNT and UCG students who were enrolled in the courses understood that they were part of a pilot study. EDSP 5560 Assistive Technology focused on training educators to effectively assess, locate, and implement assistive technology devices for students. The specialized course usually had one or two sections during the fall or spring semester.

Procedures for Data Collection

The research data included archived documents, survey results, and semi-structured interviews with stakeholders. As a part of normal coursework, during the fall and summer of 2008, UNT and UCG collected the UCG end-of-course survey and discussion board postings. The remaining data, the online learning survey, archived documents and semi-structured interviews with stakeholders, were collected during the fall of 2009.

UCG End-of-Course Survey data collection. The UCG survey was composed of 23 Likert and short answer questions. Questions 1–7 were short answer demographic questions, Questions 8–22 were Likert items, and Question 23 was a space for students to leave a comment about the course. The survey was available in Spanish and students took it online at the end of each course.

UNT and UCG collected the end-of-course survey from UCG students during the final weeks of each course through the WebCT online survey feature. The institutions requested that all UCG students complete the survey, but it was not required. The UCG end-of-course survey response rate for summer 2008 was 69% (n=18) and it was 84% (n=21) for fall 2008.

Online Learning Survey data collection. Questions 1–14 were demographic questions, Questions 15–20 and 24–32 were Likert items, and Questions 21–23 and 33–35 were open-ended questions. The data for the online learning survey was collected during the fall of 2009. Students received an email requesting their participation in the study. Participants completed the survey online through Survey Monkey and they could volunteer for an interview at that time. Contacting the UCG students was simplified through their enrollment in the final two WebCT courses. Locating the UNT students was more challenging. As contacts were attempted a year after the course, some students had completed their degrees or were not enrolled at UNT anymore. For the 5710 course, contact information for 25 out of 34 students was found. Email addresses for all the students in 5560 were found. To boost the survey response rate, non-responsive students received several email follow-up reminders to encourage them to participate and an attempt to contact students through a social networking website was performed. The online learning survey response rate for EDSP 5710 was 39% (n=20) and for EDSP 5560 was 41% (n=15).

Semi-Structured Interviews. Course stakeholders were contacted for a semi-structured interview by email and responded with their preferred method of interview (i.e. face-to-face, instant messaging, or email). Six interviews were conducted: two face-to-face interviews, two through instant messaging, and two over email. The questions were available in English. No students were available for interviews.

The UNT four-course sequence course developer and the UNT 5560 course instructor participated in face-to-face interviews. The semi-structured interview protocol was emailed to the interviewees several days before the scheduled interview to give them time to review the questions. The respondents were interviewed on the same day, one at a time, in their offices at UNT. The researcher followed the semi-structured interview protocol yet remained open to related topics. Notes and a digital voice recorder were used to document the conversations. Seventy-four minutes of face-to-face interview recordings were collected. The recordings were transcribed using a computer program called f4.

The interview responses from the 5710 and 5560 teaching assistants were collected through instant messaging. Participants received the questions in advance and were interviewed individually. The respondent met the researcher in a social networking website and began a private chat session. The researcher began the interview by greeting the respondent and then proceeded with asking the interview questions one at a time. Two and a half hours of instant messaging data were collected.

The interviews with the UCG Dean of the College of Education and the UCG program coordinator were carried out through email. Participants received an email with a brief greeting, instructions for completion, and a word of appreciation for their time. Respondents typed their feedback into the document and emailed it back. Respondents had the option of replying in English or Spanish and several did reply in Spanish. Follow-up questions were sent to the participant to clarify responses as needed. The time between sending and receiving the respondents' feedback was a few weeks and 10 pages were collected.

Archived documents. The archived documents were collected by requesting permission from the IRB, the course instructors, and other university personnel as needed, as well as using online databases. Archived data from 2008 included the syllabi, conference papers, and presentations, WebCT discussion postings, grant proposal, course averages by groups, course assignments, quiz and exam items, completion rates, and the UCG end-of-course survey.

Reliability and Validity

UCG End-of-Course Survey instrument validity. An exploratory factor analysis and a hierarchical cluster analysis revealed two factors: university communication and support, and quality of course content. Cronbach *alpha* reliability coefficients were computed. The entire scale had an internal consistency reliability of .76, (n=25). According to DeVellis (2003), all *alphas* fell into the "respectable" range. Items 9, 10, 13, 14, 16, 19_R, and 20 were omitted due to low factor loadings or that their removal increased Cronbach's *alpha*.

Online Learning Survey instrument validity. This survey was adapted from Ruhe's dissertation (2002), which was originally constructed by a committee of project associates from The University of British Columbia based on Bates's (1995) ACTION framework. It was pilot-tested and revised. Ruhe (2002) used 90 quantitative items and 6 short answer questions in her dissertation. For this research, 35 questions were used,

29 quantitative items and 6 open ended questions. In addition, eight questions about international computer-mediated communication were created. The questionnaire items are from Ruhe (2002) unless indicated otherwise and measure technology skills (adapted from Gold 1997), course access, benefits, communication (researcher created), and multicultural awareness (adapted from Guyton and Wesche 2005). To ensure the integrity of the survey translation from English to Spanish, the survey was submitted to a double review process. First, the instrument was professionally translated; then, a UCG stakeholder reviewed the translation and made corrections. The instrument was available online in English and Spanish.

An exploratory factor analysis and a hierarchical cluster analysis found three factors: student openness to international partnerships, students' value of communication with the partner university, and technology skills. Cronbach *alpha* reliability coefficients were computed. The entire scale had an internal consistency reliability of .81, (22). According to DeVellis (2003), all *alphas* fell into the *respectable* to *very good* range. Item 32 was omitted due to low factor loading.

Procedures for Data Analysis

Applying the framework to the data using a mixed methodology. Both quantitative and qualitative measures were employed through a framework called the unfolding model (Ruhe and Zumbo, 2009), which is based on Messick's (1989) work on validity. Ruhe and Zumbo (2009) created the unfolding model to answer the call in literature for a professional evaluation model that would be adaptable to guide studies of merit and worth for distance learning programs. The unfolding model utilizes mixed-methods to collect and analyze data along the four facets central to the framework: scientific evidence, relevance/cost-benefit, underlying values, and unintended consequences. The four central framework facets are further divided into scientific basis and consequential basis. It is important to point out that Ruhe and Zumbo (2009) do not define science as controlled experiments but as a rigorous examination of how social realities emerge, function, and affect individuals and organizations. This model was chosen because it can be used for formative evaluation, it provided structure, captured the perspectives of a wider audience, and allowed for overlapping data. The two courses, EDSP 5710 and EDSP 5560, were treated as individual cases by performing an in-depth study of the components of each course.

We applied the model to our data through a process of "cycling through" the framework, as Ruhe (2002: 101) coined it. By holding up the unfolding model framework against the data like an overlay transparency and through a continual process of sense making, the primary researcher decided where and how to place the data within the four components. The data were analyzed by searching for gaps between the ideal and the real; that is, the difference between what course instructors hoped would happen and what actually happened. The primary researcher examined whether there are conflicting viewpoints between the underlying value of multicultural awareness and actual course implementation (Ruhe and Zumbo, 2009). To do this; first, the data were analyzed and compared with all the coded archived documents, open-ended survey responses, and Likert survey subscales. Second, the coded semi-structured interview transcripts of respondents were examined. The goal was to use all of the data to see how the course worked as a system in order to determine how well the course operated. Ruhe and Zumbo (2009) refer to this as determining the goodness of "fit" within the unfolding model framework (p. 195). The data were analyzed for "fit": between the multicultural components and the course.

*Quantitative data.*The quantitative data were analyzed using the statistical software, SPSS. We followed this process to evaluate both surveys. For each survey, descriptive analyses were employed on the demographic data, along with an EFA and hierarchical cluster analysis. Cronbach's alpha reliabilities were computed and scaled scores were created (DeVellis 2003).

Qualitative data. Qualitative analyses began deductively, as the quadrants of the unfolding model were used as *a priori* coding categories. Inductive analysis began after the bulk of the data had been sorted. In this stage of the analysis, the primary researcher remained open and flexible to themes as they emerged from the data within and among the categories. The content analyses followed Patton's (2002) suggested stages of identifying, coding, categorizing, classifying, and labeling the primary patterns in the data. This process required that the principal researcher continually check the data against their own perspectives in order to make sense of the evidence. Direct quotations from interviewees were selected to enhance our findings with thick/rich description. The qualitative software, MAX-QDA 2007, was used to organize and assist with the specific units of the analysis.

Results

The evidence resulting from our four central questions revealed these findings for each course.

1. What is the evidence of multicultural awareness in these two special education courses?

Concerning EDSP 5710, as demonstrated through the course textbook, assignments, standards, and other course materials, multicultural awareness references and objectives were woven throughout this course. Data from the online learning survey revealed several findings. Students most frequently referenced that multicultural education is about acquiring a different perspective. This is reflected in student C_34's statement that multicultural education is "Inviting students from other cultures to help our understanding of the subject matter." Also, in student C_52's reflection that "El intercambio de culturas, nuevas formas de percibir otra cultura, estimula la tolerancia y el respeto a las diferencias." [The intercultural exchange and formation of new form of perception of other cultures stimulate tolerance and respect for differences. (trans. Tcherepashenets)]. Students referenced the exchange of information as a component of multicultural awareness. This is seen in the response from student C_3:

> In the context of having a partner university in Ecuador, I think the term multicultural education would relate to students from both US and Ecuador exchanging information pertaining to their own culture's attitudes toward education. This would include subjects such as expectations of teachers and administrators, parents attitudes toward education, funding available for SPED programs, gender biases in recommended curriculum, incorporation of cultural diversity studies into lessons taught…I could go on forever.

Students were evenly divided on whether they viewed the course materials as enhancing their multicultural awareness. Those in favor mostly just responded with "yes"; however, the response from student C_20 described that the course allowed her to know the US perspective. "Si, porque me permitieron conocer qué se hace en EE su normativa, etc." [Yes, because it allowed me to understand what is going on in the United States, its norms, etc. (trans. Tcherepashenets)] On the other hand, student C_3 reflected the opposite point of view, "Cultural diversity has been an underlying theme in most of my courses. As I stated before, I did not communicate directly with any of the Ecuadorian students so I did not gain any specific awareness of their culture."

Over half of students responded that the course had influenced their lesson design regarding the inclusion of various cultures. Student C_12 reflected this thought about being more creative in her lesson design:

> Sí ha influenciado en ser más creativa pero orientado a involucrar varias culturas no relacionadas con lo étnico, sino con grupos sociales, por ejemplo: grupos de maestros, grupos de estudiantes jóvenes, grupos de estudiantes mayores. [Yes, it influenced me in becoming more creative through inclusion of different cultures, not in the ethnic sense, but in the social sense, for example: a group of teachers, groups of tradition students, group of adult students" (trans. Tcherepashenets)].

Student C_3 emphasizes that the course affected her beliefs about the importance of valuing other cultures:

> I feel it is critical to include studies of different cultures throughout the student's educational process. I feel 'Americans' have become quite arrogant and forgetful of their multi-cultural backgrounds. Cultural sensitivity applies to different sectors of the US population as well as those who reside in different countries.

In contrast, student C_29 was an example of those who felt that, for a variety of reasons (e.g. translation issues, lack of contact with student/instructor, or the course adaptations) the course did not influence their multicultural awareness.

> No siento que el curso ha tenido impacto en una mirada cultural diferente para mí, puesto que no hemos tenido contacto con otros estudiantes. En cuanto a lo intecultural del intercambio con los profesores, lo unico que me ha afectado hace referencia a la calidad de las traducciones y de alguna manera a los contenidos, pusto que estaba completamente realizado para público y realidad de USA. [I do not feel that the course had an impact on my vision of the cultura, because we did not have contact

> with other students. As for the intercultural experience and exchange with professors, the only thing that had an impact was the poor quality of translation and some content, which was created completely for the US audience. (trans. Tcherepashenets)]

Findings from the UCG end-of-course survey suggested that students perceived the support from the universities as *very good*. That thought was also echoed in the open-ended responses.

For EDSP 5560, we distinguished nine multicultural awareness references in the course assignments. Data from the online learning survey revealed several findings. Students most frequently referenced multicultural education as acquiring a different perspective. This is reflected in student C_24's statement that multicultural awareness is "Learning about the different ways that students from different cultures learn. What they learn, their experiences, etc." Also, it is seen in student C_9's response "Cambio de experiencias desde las culturas personales, que no necesariamente se dan por estar en diferentes paises, pues dentro del mismo pais, ciudad o barrio se puede dar." [Change in perception of personal cultures is not necessarily related to the staying in different cultures, it can happen within one culture, city or district. (trans. Tcherepashenets)]. The majority of students viewed the course materials as enhancing their multicultural awareness. Those in favor, mostly just responded with "yes;" however, the response from student C_10 describes in more detail how the course influenced her awareness. "Sí la aumentó, porque yo conocía poco y porque en nuestro pais no hay todos los materiales que he conocido a través del curso." [Yes, the course increased it, because I knew little and in our country there is no acces to materials, with which I became familiar during this course. (trans. Tcherepashenets)]. In contrast, student C_29 sheds light on the translation issue, "Creo que no tener buena traducción impide un buen aprendizaje." [I believe that the lack of a good translation impedes learning. (trans. Tcherepashenets)].

More than half of students responded that the course made a difference or influenced their lesson design. Student-perceived openness to international partnerships was found to be between the range of *agree* to *strongly agree*. Findings from the UCG end-of -course survey suggested that students perceived the support from the university as *average* to *very good*. Overall, the most frequent theme for UCG support was that students received sufficient assistance. Concerning UNT support, students responded that the courses offered a new source of information. In addition, stakeholder_15 summarized the multicultural awareness goals of the courses nicely. She was asked: "What did you hope they [students] would walk away with after this was all over?" Her answer was:

> I wanted them to become what I consider, global educators. Educators that don't necessarily understand all that components and aspects of every single culture on the face of the world, but at least understand that there are people out there that are very similar yet different from us and have very special needs. And understand and address those needs when they are dealing with children.

2. What is the cost-benefit analysis of the course for the institutions?

The University of North Texas Global Initiatives Grant for $18,000 funded the four-course sequence of which 5710 and 5560 were the first two courses. The grant provided for the travel expenses for UNT faculty and staff to conduct the face-to-face training in Guayaquil, Ecuador, the translation and development of the two courses, and a bilingual grader for each course. UNT benefited by strengthening an international partnership and developing bilingual courses to offer to other Spanish speaking universities. During our interview, UNT stakeholder_15 acknowledged the additional advantage that as the world shrinks due to online learning, bilingual course offerings will become more prevalent:

> But I think it is a good movement. I think it is something we're going to have to look at in the future as far as languages go. I think that those day is about to really shine on us and we're going to need to consider this.

UCG's cost was $150 per course paid to the UCG program coordinator. UCG benefited from the courses by receiving the courses at no cost from UNT. Overall, the stakeholder consensus was that the courses went well, in addition, a number of respondents selected "There was group of students that welcomed assistive technology" and "Instructor gained from the experience." The statement from UNT stakeholder_28 expresses the positive experience, "I love the courses; I love how they turned out and I loved being a part of it. It was an awesome project and I wish to do more things like this in the near future." According to the email interview with UCG stakeholder_33, she was able to see two groups form, one which embraced assistive technology and another that thought it was out of reach.

> Sin embargo a lo largo del curso, podemos decir que se vio dos grupos: aquellos que acogieron positivamente la propuesta de la tecnología asistencial como algo posible e importante de promover, y aquellos que vieron que la tecnología asistencial es algo distante y caro para nosotros. [Nevertherless, during the course, two groups emerged: one, which embraced the use of assistive technology and its affordances as something possible and important to use, and others who viewed assistive technology as something distant and expensive for us. Trans. Tcherepashenets.]

One of the most advantageous unintended positive outcomes of 5560 was that during the first few weeks into the course a UCG student was inspired to organize and host an assistive technology forum at UCG. Experience gained from this partnership has led UCG to begin plans to develop its own master's degree in special education.

3. How was the underlying course goal of enhancing students' multicultural awareness implemented in the course?

For EDSP 5710, in general, UNT did make a conscious effort to adapt the courses. The instructor clearly stated the expectation that UNT and UCG students should communicate in several ways. First, the syllabus listed that students should post to the discussion board. Second, the instructor and teaching assistant posted several messages to the discussion board introducing UCG to UNT and explained where to find online translators. They also provided general encouragement for students to collaborate. In this course, the teaching assistant was not assigned to translate course postings. As a result, the UNT students mainly posted to other UNT students and the UCG students did the same, although a few students did reach across international lines. Periodically, the instructor or teaching assistant would attempt to bridge the discussion board divide with a bilingual post. The instructor was diligent to welcome students to the class after they had posted an introductory message; as a result, some of the students' replies indicated that they perceived themselves as valuable contributors to the class. Some multicultural awareness messages appeared in the discussion board.

For 5560, overall, UNT did make a concerted effort to adapt the courses. The expectation that UNT and UCG students should communicate was clearly stated in several ways. First, the instructor wrote in the 5560 syllabus, that the discussion board postings were required as components of two course assignments. Also, the instructor included specific instructions in the rubric on writing an appropriate post. Students were to write one initial post and one response to a post on the discussion board for two assignments. Second, the instructor incorporated nine discussion board prompts in English and Spanish. The instructor's goal was to create a welcoming atmosphere within the course. Third, the instructor asked the teaching assistant to serve as a translator for the discussion board in order to assist with bridging the language barrier between students. As a result, the UNT and UCG students communicated more frequently. Periodically, the teaching assistant would translate a lengthy post from one of the students to enhance communication between the groups. In some ways, 5560 opened up a completely new world for the UCG students and they were excited about the possibilities of assistive technology within their own schools.

4. What are the unintended positive or negative consequences of the course design and implementation?

The inclusion of multicultural awareness appears to have taken place in all areas of the course materials, which would suggest that the course goals and implementation aligned. When students were asked to describe their own view of multicultural education, they appeared to understand the meaning of the word and a large majority viewed 5560 as increasing their multicultural awareness. The data suggest that for some students, 5560 did increase their multicultural awareness and influenced their inclusion of various cultures in the classroom. In addition, the frequent instances of intercultural communication on the discussion board aligned with the grant's goal of enhancing teachers' multicultural awareness. In my interview, the 5560 instructor conveyed the multicultural awareness theme as well. One of the instructors, for example, observed:

> First of all, I wanted them [students] to know about assistive technology and be able to implement it in their classrooms, all the kids, the Americans as well as the Ecuadorian students. That was the primary thing… Anything in addition to that in terms of multicultural awareness, appreciation for the systems of other countries, that's kind of like icing on the cake. That was, just that much better, that much more.

Our analysis revealed that translation issues for both courses were a major challenge. The language barrier emerged as a concern with both UNT and

UCG stakeholders. By email, UCG Stakeholder_55 shared that the problems with the translation stem from sentence structure, vocabulary, and confusing passages. With UCG, the theme manifested itself as appearing that the instructor gave little feedback. Meanwhile at UNT, the instructor felt the same distress as she attempted to respond yet lacked the skills. She explained the situation in an email interview:

> So they (UCG) would write me, but I was of no help really because I'm not bilingual. What I would do is I would paste their comments into one of these google translators, babelfish, or something to figure out exactly what it is that they were saying and then I would try to answer them to the best of my knowledge.

Specific issues concerning 5560 were that UCG students frequently referenced improving the professor/teaching assistant feedback as their highest recommendation to improve the course.

Students most frequently responded that flexibility and limited driving were the greatest benefits of online learning. Student C_34 emphasized that thought and wrote "Was able to do it on my time." Also, student C_ 26 responded that being able to work at home benefits her by "not having to go to campus." And in student C_ 32's feedback on her technical difficulties submitting assignments. Students appeared not to have any major unexpected issues with course. Besides responding with "No," "Translation issues" were the next highest concern. Student C_33 described that the issues with the translations involved, "la forma de traduccion de ciertos documentos." [the form of translation of certain material. (trans. Tcherepashenets)]

The UNT stakeholders focused on recommending improvements with the logistics of course delivery such as enrollment and conducting onsite meetings while the UCG stakeholders recommended improving the translation through a double-review process and including more instructor feedback. The stakeholders' recommended that the framework be strengthened for non-bilingual instructors and students to be able to communicate more freely:

> Dr. E and I are looking at options for offering the online classes to other Spanish speaking countries. Dr. S is offering his support for providing the courses to other South American countries. We will take your suggestions and update the classes before they are offered again. The review of the classes has helped us to better understand the needs of the UCG students. UCG students require more direct instruction from the instructor. In the future we will staff the classes with instructors who have

> better skills in Spanish. The face-to-face meeting time was important. We will attempt in increase face-to-face meetings in the future (Stakeholder_15, personal communication, December 16, 2009).

The respondents' recommended that the framework be strengthened for instructors and students to be able to communicate. One suggestion from stakeholder_20 is to involve more than one translator. Also, stakeholder_54 proposed incorporating more background information on Ecuador for UNT students to facilitate their understanding. Throughout all our data, we consistently found UCG offering thanks and gratitude to UNT.

In conclusion, both 5710 and 5560 took an innovative approach to bridging the digital divide between two culturally different groups. Our findings suggest that the courses had many strengths, as well as a few weaknesses. We present the discussion of those components next.

Discussion and Conclusions

Researchers have conducted program evaluations of distance learning programs in the US; still, very little research exists on the evaluation of international online special education programs. Although one study does not provide all the answers, our findings contribute to the literature by evaluating two bilingual, online special education courses, when previous research was limited or nonexistent. The internationalization of distance learning in special education is at a pivotal point in expansion. As mentioned earlier, with the global need for more special educators, many institutions are turning to distance learning to increase the offering of special education resources. One concern is that institutions may prematurely expand their courses internationally before the proper bilingual framework is in place. The broader implications of our research apply to the development of international distance education programs. Our findings reveal that institutions need to take several issues into consideration while developing a telecollaborative, bilingual special education program. First, institutions need to survey the availability of course materials in the language of the receiving country and consider these questions: Where will the course developers find the materials? How will the course developers adapt the materials? Is there a digital divide? Second, the course development team must establish a support system for the international students. Institutions can do this by conducting face-to-face training sessions, providing an onsite liaison, and staying in close contact with the receiving university to monitor the students' needs. This may require that the US instructors modify their own teaching style to meet the needs of the international students. Third, institutions should encourage intercultural dialogue and provide an avenue for all students to feel connected in the course, even if they do not speak the same language. This means that an institution needs to provide multiple translators and, ideally, instructors with a working knowledge of the language. Last, the course translation should reflect the language and culture of the receiving international group. Converting the course from language one (L1) to language two (L2) is one of the most challenging aspects of course development because direct translation does not exist for all words and meaning can be relative to a geographical location. Institutions can improve the translation by treating the L2 coursework as a completely separate course, instead of as only a twin of the L1 course. The L2 coursework should be customized to the learning needs, culture, and language of the L2 students. Extra care should be taken to accurately translate high priority items such as quizzes and exams due to the sensitive nature of students' grades. Developing an international course is labor intensive and requires several drafts before implementation. Institutions should conduct a double-review process in which a native speaker from the area reviews the translation before the course is finalized. Even with a stringent review process, both institutions need to understand that translation issues will arise and institutions need a protocol for correcting the errors.

Diverse cultures can be integrated by investigating ways that international students can make learning personal by putting it in their own context. What is deemed an appropriate time frame for receiving feedback for questions and graded assignments, for example, can vary by culture. Institutions will benefit by clarifying student and instructor expectations in this regard. If the hosting institution makes a concerted effort to collaborate with the receiving institution, then even when cultural differences emerge, the institutions will be able to work through the difficulty. Above all, the receiving institution will appreciate any efforts to offer an international course, even if the courses are not perfect.

In sum, before embarking on an international special education distance learning program, institutions would be wise to consider what resources are available in the language, what type of student support system to put in place, how to enhance telecollaboration, and how to translate the courses to reflect the international students' language and culture. If implemented correctly, more international special educators will receive much-needed training that is relevant and meaningful. The findings from this research project further the development and refinement of cross-cultural online education for special educators by evaluating a pioneering effort to prepare teachers to live in a globally connected world. Once thought an aspiration for the future, international sharing of special education programs is a social justice issue and must become a reality if institutions are to meet the acute global shortage of special educators. International telecollaboration is vital in visualizing special educator training in the future.

Indeed, the findings are significant to the development of collaborative international academic efforts regardless of discipline, in order that those of differing social, economic, and cultural backgrounds may receive equal educational treatment and outcomes. Through such ventures, skills that are pertinent to one's discipline will be enhanced while learners grow in multicultural understanding and global citizenship.

References

Banks, J. A. (2001) Diversity, group identity, and citizenship education in a global age. *Educational Researcher* 37(3): 129–139.

Bates, A.W. (1995) *Technology, open learning and distance education.* London: Routledge.

Binner/Falconer, K., and Lignugaris/Kraft, B. (2002) A qualitative analysis of the benefits and limitations of using two-way conferencing technology to supervise preservice teachers in remote locations. *Teacher Education and Special Education, 25*: 368–384.

Bore, J. (2008) Perceptions of graduate students on the use of web-based instruction in special education personnel preparation. *Teacher Education and Special Education, 31*(1): 1–11.

Center on Personnel Studies in Special Education. (2004, February) *An insufficient supply and a growing demand for qualified special education personnel*. Retrieved 12 October 2008 from <http://www.coe.ufl.edu/copsse/docs/PB-22/1/PB-22.pdf/>.

Cerny, M., and Heines, J. (2001) Evaluating distance education across twelve time zones. *Technological Horizons in Education Journal,* 28(7): 18–26.

Coutinho, M. and Oswald, D. (2000) Disproportionate representation in special education: A synthesis and recommendations. *Journal of Child and Family Studies,* 9(2): 135–156.

Cummins, J. (1989) A theoretical framework for bilingual special education. *Exceptional Children, 56*: 24–30.

DeVellis, R. (2003) *Scale development* (2nd ed.). Thousand Oaks: Sage Publications, Inc.

Fore, C., Martin, C. and Bender, W. N. (2002) Teacher Burnout in Special Education: The Causes and The Recommended Solutions. *The High School Journal* 86(1): 36–44. The University of North Carolina Press. Retrieved 16 March 2013 from Project MUSE database.

Gold, B. (1997) *A formative evaluation of a distance learning program at CSUDH.* (Doctoral dissertation, Pepperdine University, 1997). Retrieved 4 September 2009 from Dissertations and Theses: A and I. (Publication No. AAT 9814563).

Part II: Web 2.0 and Social Change

Derya Kulavuz-Onal and Camilla Vásquez

Opening up Worlds: Intercultural Exchanges through Telecollaboration

Introduction

Over the last few decades, developments in communication and information technologies have reshaped our understanding of the nature of communities. Traditionally, communities were collectives that were bound by some type of spatial constraints: most typically a group of persons living in close proximity to one another and inhabiting a particular geographic region. However, with the arrival of the internet, and with the subsequent evolution of Web 2.0 (an era associated with expanded opportunities for interactivity online), increasing numbers of people are joining online communities. These types of online communities emerge with members who – although often geographically dispersed – share some common interests, activities or affinities. Indeed, some contemporary scholars of new media argue that our membership in such "interest" or "taste" groups is becoming just as important to our online social identities as is our membership in the more traditional categories of regional origin, religious affiliation, occupation, etc. (e.g., boyd & Ellison, 2007; Liu, 2007; Liu, Maes & Davenport, 2006).

In this chapter, we discuss some of the activities of a global online community that has been in existence for ten years. The community in focus is a multi-site, community of practice (CoP) of English Language teachers whose interests and practices center on learning and using new technologies for professional and personal purposes. The group's members are found in dozens of cities and countries around the world, and each member inevitably interacts with members from numerous other countries. In addition, membership in this community entails engagement with various forms of telecollaboration, which we will describe below. Then, we focus on and discuss in detail one specific case of intercultural telecollaboration between two English teachers within this community that emerged from a larger netnography (Kozinets, 2010) of this online community (Kulavuz-Onal, 2013), for which we collected data through online participant observation, interviews with community members, as well as archived communications. Our focus in this chapter is to describe how telecollaboration provides members the opportunity for meaningful intercultural exchanges, and to illustrate one instance which extended into two different classroom contexts with limited technology on two different continents. We argue for the potential of using such technologies to enhance intercultural understanding, which we believe is essential in the promotion of social justice.

Webheads in Action Online Community of Practice

The roots of Webheads in Action (WiA) Online Community of Practice were planted in a free online workshop offered as part of the Electronic Village Online (EVO) sessions organized by Teaching English to Speakers of Other Languages (TESOL) Computer-Assisted Language Learning Interest Section (CALL-IS) in 2002 (Stevens, 2007). This initial workshop was offered and moderated by Vance Stevens, and was called *Webheads in Action: Community Formation Online and Its Role in Language Learning.* The participants in this workshop continued to keep in touch even after the workshop ended, and in the following years, the name of the workshop became the community's name, and the members began calling themselves Webheads.

Webheads are English language teachers or professionals from all around the world. Their primary interest and practice is to explore and exchange ideas on the pedagogical uses of web-based technologies, and use these technologies to interact and collaborate with each other, as well as to integrate them in their own classroom teaching practices. They develop their collective practice through collaborative and voluntary activities/events that they themselves organize entirely online, and archive at several venues on the Internet. These archived wikis, blogs, recordings, etc., in turn, serve as open-source, free professional development resources for not only the members, but also for any English language teachers worldwide who are interested in using them.

Although, at the beginning, the WiA community was mostly bounded by a Yahoo Group email list, the evonline2002, which was created for the initial online workshop that brought the members together for the first time, it has gradually emerged as an online CoP that spreads over multiple venues as members carry out and archive group activities, as well as interact, through various platforms, social media, and computer-mediated communication (CMC) technologies. As a result, the WiA online community does no longer exist or is bounded by only one website or a single online platform. However, as a community, they continue to keep the initial Yahoo Group email list (i.e., evonline2002) as their primary interaction platform. As of September 12, 2012, there are 1,005 members registered with the group's email list. The members of the community are located in various countries all over the world, spanning from Japan to Canada and Russia to Argentina, and can be found on all the seven continents.

In addition, Webheads has a portal which includes links to their activities and venues (<http://webheads.info>). On this website, the community's mission is described as "an online community of practice of teachers and educators practicing peace and professional development through Web 2.0 and computer-mediated communication." It can be understood from this description that world peace and social justice as well as members' professional development are important concerns for Webheads. This ethos is explicitly displayed, fostered, and nurtured in their activities, values, and organizational structure, as we illustrate below.

Webheads, in line with the principles of Open Access movement, support the belief that knowledge should be freely accessible. Consequently, they organize activities, develop their collective practice, and interact using either open source technologies such as Yahoo Groups, Twitter, wikis, and blogs, or make their activities publicly available through their non-profit sponsors (e.g. the free virtual room through Blackboard Collaborate provided by LearningTimes.net). In return, they offer their activities and archives (e.g. synchronous session recordings, chat logs, wikis, etc.) freely online. For example, every Sunday interested members gather together for *Learning2gether* meetings (<http://learning2gether.pbworks.com>), which are usually held through Webheads' virtual room on LearningTimes.net, or WIZIQ (a free web-conferencing platform). Each session is announced through the Yahoo Group email list, and Twitter, and is recorded. The recordings then are made available through this public wiki and <http://learning 2gether.posterous.com> (a free blogging tool). Another free activity that they collaboratively and voluntarily organize is the *Becoming a Webhead* (BaW) online workshop series that is offered annually, as part of the EVO sessions. These annual workshops serve as the beginning steps for the new members to become oriented to Webheads' culture, activities, and practice. Having been carried out annually since 2005, these 5-week free workshops are organized similar to an interactive online course that uses open source CMC technologies (e.g. Yahoo Messenger and Skype), web 2.0 tools (e.g. wikis through <pbworks. com>, blogs through Blogger, online forums through Proboards, etc.) and social media (e.g. Twitter). As such, both participation in their activities and access to the materials and resources (e.g. see BaW2011 wiki at <http://baw2011.pbworks.com>) created by the community are freely and equally accessible by anybody with an internet connection. What makes these workshops unique is the fact that the content (from syllabus to activities), which is delivered via an open-source wiki, is co-constructed by the members through voluntary efforts; they are not organized or moderated by a community leader or community founder. Moreover, each week, one or two guest speakers, who are also Webheads, volunteer to deliver a presentation on the topic of the week over Blackboard Collaborate. Using free online bulletin boards, the participants discuss their ideas about the weekly topics and readings. At the end of the workshop, the first-time participants graduate as "Webheads," and are invited to register to the main Yahoo Group email list (i.e., evonline2002). Through this main list – and by participating in the community's subsequent activities – new members are able to keep in touch with the other new and old-time members.

Another characteristic of Webheads is that they value each other's experience equally and they believe that every member has something to share. It is for this reason that it is difficult to locate a single "community leader" in this CoP. For example, membership roles are loosely defined and while one member is active in one activity, s/he may be less active in another activity. All members take initiative for carrying out at least some of the group's activities. For example, one member voluntarily creates a blog almost every year for those who attend the annual TESOL Convention in the US. This person shares observations, experiences and reflections about the conference sessions on the blog, so that the other members who were unable to attend can gain more of an "insider" view of the convention. In addition, another major activity organized by Webheads – i.e., the *Learning2gether* weekly meetings – also embody the group's belief that "everybody has something to share". Vance Stevens, the community co-founder, coordinates the *Learning2gether* wiki, however, any volunteers who want to present a technology-related teaching practice, or a study, or to simply lead a discussion that would be of interest to the community members, are invited to add their names and session information to the slots available on the wiki. There is no previous requirement for presenting during a *Learning2gether* session, other than having a desire to share one's ideas and experiences with others. In addition, the selection of moderators for the annual BaW workshops also illustrates how members value each other's experience and expertise, regardless of how recently they have joined the community. In these workshops, active participants who exhibit "Webheads values" (e.g. a willingness to volunteer, help and share) are invited to become moderators for the next year's workshop. This gives an opportunity for those members to have an e-moderating experience as well as to further develop their professional expertise.

Intercultural dialogue occurs naturally and frequently in Webheads' interactions with each other because of its globally-distributed nature. For example, one convention among Webheads is to sign off on their emails by including information about their location. For instance, this is sometimes made explicit by closing their message with the name of a city or country (e.g., "*E-kisses from La Plata*", "*Bye from Croatia*", "*Hugs from Rosario*"), or more implicitly with a conventional closing written in a language other than English (e.g., "*Beijinhos*", "*Beijos*"). Moreover, when they hear about breaking news in a country where other Webheads are living, they send messages to those individuals through the evonline2002 email list. For example, during two major events that were globally broadcasted in 2011 (i.e., the Egyptian Revolution and the Japanese earthquake), members sent messages of concern and good wishes to other members who were directly affected by these events. In that sense, their community membership and attachment to each other not only facilitates empathic relationships but also enables them to be more alert and aware of current news in each other's countries, as well as to further develop their intercultural, interpersonal relationships, regardless of where they are located in the world.

Another common practice among Webheads is telecollaboration that occur in many ways. For example, one form of telecollaboration happens when they work together, through online communication, in order to put in a proposal and a presentation for various face-to-face and online conferences such as TESOL, which caters to English language teachers' professional development. Additionally, they contribute to each others' teaching and courses synchronously, or asynchronously. For example, it is not uncommon for a webhead to become a guest speaker in another webhead's class in another country through a web-conferencing platform. Also, whenever they conduct a technology-infused project with their students, webheads showcase students' projects to other webheads, and solicit their comments or contributions to these projects. Another common way is to collaborate together on cross-cultural projects including their students, in which students share cross-cultural information with another teacher (who is also a webhead) and their students through online means.

In the next section, we illustrate one specific intercultural telecollaboration that took place between two Webheads: Ayat, an English teacher in Egypt, and Maria, an English teacher in Argentina. This specific case emerged through our netnographic fieldwork for the larger study we conducted through January – November 2011. Specifically, we describe how – through their engagement in the WiA community – they formed a professional relationship and friendship that led to specific telecollaborative projects among them and their students. We illustrate in detail how their relationship evolved into a series of telecollaborative exchanges, which enabled their young students to learn about each other's cultures. We conclude by discussing how this project fostered intercultural understanding, and specific ways in which these types of community practices embody principles of social justice. In the discussion that follows, we draw on data from our participant observation fieldnotes, archived materials, individual interviews with the teachers, as well as a group interview with both teachers. In keeping with the principles of netnography (e.g., Kozinets, 2010), all data were collected online.

The ArgentEgypt Telecollaboration Project

The ArgentEgypt project took place between two English as a Foreign Language (EFL) teachers who met during the BaW workshop in January 2011: Ayat, a teacher of 10–11-year-olds at a private school in a large city in Egypt, and Maria, an English teacher at a private institution in a large city in Argentina.

Having met through the BaW workshop, both teachers were affected by their interactions with the Webheads community similarly. From her earliest interactions with BaW, Ayat felt as though her personal and professional life was transformed. In an interview, she explained that after participating in BaW, she felt that "doors being opened and worlds being opened to me till the sky" as she learned new skills and tools, gained confidence in her teaching abilities with technology, as well as making friends around the world and learning first-hand about places she knew very little about. Considering that she is from Egypt, and the Egyptian Revolution took place during BaW2011 when Ayat had no internet connection for a week, her comments about how this online community and her interactions with them through Internet opened up worlds for her are particularly striking.

> And personally of course, it has changed me a lot. I have lots of friends now, which I didn't use to have. And from lots of countries worldwide, I feel that I can travel everywhere now, and I have known about many different countries, which I haven't heard about before. [...] So, I have gained lots of friendships, and even not just friends, but I have some close friends now, which I have never had in real life here, believe it or not. *I feel more free now, I open the internet and as if I open a world in which I can fly freely and be in and pick whatever I want. Once I close the internet, then this world just closes and I'm back to my real life.* (Ayat)

For Ayat, online is place where horizons expand, where possibilities for learning and interaction are infinite, and where she feels freer to make choices. This reality contrasts with her "real life," or her offline reality, where her "world closes" back up again. Similarly, even though Maria had been involved in other EVO sessions before, she describes her experiences in BaW as a "great jump" that "marked the difference" and that "opened my mind to technology" (Maria).

The Beginnings of a Cross-cultural Partnership

Once the 5-week BaW workshop ended, both teachers joined the WIA list serve and officially became "Webheads." As they continued chatting online about professional matters, they got to know one another and became closer. As Maria explains, this online contact evolved into not only a professional relationship but also an intercultural partnership once they began sharing more of their personal and professional lives and cultures with each other in constant interaction:

> At the very beginning I guess it was professional contact. We started talking and then after talking a lot about profession, we started our friendship [...] I asked Ayat a lot of questions [...] I always asked her about her culture, about her religion, about her – not her private life, but her life in general in Egypt, her culture, in fact. So I guess I started the relationship from there, then she started asking about my culture, my life. (Maria)

Moreover, Ayat had never before experienced building a friendship with someone from another country. This enabled her to experience a closer cultural connection to Maria than she does with many others who actually share her own culture:

> That's one of the points that really made me feel very close to Maria or very attached to her when she told me that she had lots of friends in Egypt and she knows a lot about our country, about our religion... So I was like astonished, I didn't have such a kind of friendship or even context from other countries...So it was the first place where I could have friends from many other countries [...] when I talked to Maria, I started to feel that she is very close to me, to my culture, to my country [...] Sometimes I even felt that she was even more friendly than my friends here, my real, face-to-face friends here. (Ayat)

Learning Culture through Student-Teacher Skype Interviews

In their interactions through these online chats, Ayat and Maria learned more about each other's educational system, teaching beliefs and styles, and cultures. Because they both were eager to experiment with the technology tools (e.g. wikis, blogs, VoIPs (Voice over Internet Protocols), digital storytelling tools, and social media) that they learned in BaW in creative ways, they brainstormed about possibilities for collaborating together. Considering their own limited-technology contexts, it appealed to both of them to first integrate a Skype project for several reasons: It was user-friendly for both themselves and their students; it was one tool that was available in their own teaching contexts with limited technology; and such a VoIP could actually give their young students confidence to speak in English when they speak with a real person from another part of the world. At one particular moment, Ayat made a connection between a lesson she was preparing for her students and Maria, in Argentina. Ayat's class in Egypt was studying a unit on global warming. Their textbook mentioned rainforests, and that most rainforests were located in South America. Ayat asked her students "Can you name some countries in South America?" Her students responded by naming Brazil, Chile, and Argentina. She then asked them "What is Argentina famous for?" They responded with the name of a few famous football players. She asked them "What ELSE is Argentina famous for? Other than football?" Below she explains her vision of how she could make this lesson come alive for her students:

> And then I had a lesson about South America, and the cities there, the rain forests, and so on. Then we mentioned Argentina, and students became very excited because they know the football players there and so on. So I used these points to start an interview with Maria. I told them what about interviewing a teacher from Argentina and asking her whatever you want to know about that country. (*"What do you think about interviewing a person from Argentina? Would you have any questions for him/her? Would you like to know more about Argentina?"*) And they were very excited, they prepared questions. And then they prepared questions and they couldn't imagine how we are going to interview her, and how we are going to see her life and so on. (Ayat)

Most of Ayat's students were unfamiliar with Skype, and this was going to be the first time that they connect to Internet and speak with a person from another part of the world, in class. Therefore, they weren't able to imagine how the interview would unfold, or that they would be able to actually see this teacher from Argentina. Also, Ayat did not have many expectations at the beginning, since this was going to be the first trial. Therefore, some students prepared their questions, while others remained somewhat skeptical about the prospect. But any initial reservations quickly shifted to excitement and enthusiasm, as students entered the computer lab and were able to see Maria's face on the screen.

> And once we entered the computer lab, once they saw Maria live, and they talked to her and she could see and talk with them, they were very very excited. Each one of them got [to] hold the mic to ask her the question, and those who were not interested before the interview actually regretted it so much. Because they don't have the question, they didn't prepare the question, they didn't share in this interview. (Ayat)

After the success of this experience, the teachers decided to reverse roles, and Maria had her students prepare questions for Ayat. Maria explained that Argentine students are curious to learn more about Egypt from the time they were first introduced to ancient Egypt civilization in primary school. Her students were especially excited about speaking with a "cultural insider," (from a culture which, for them, seemed to be closed to modern world) and someone who was physically located in a different part of the world. Maria's students asked Ayat questions such as "What is Egypt like? How big is Cairo? What can I see at the pyramids?" Even more interesting was that both students were also interested in making an interpersonal connection with the teachers they interviewed, asking the teachers not only questions about their countries, but also more personal questions such as "When is your birthday? What is your favorite movie? Who is your favorite actor?" In doing so, students were able to learn that culture is more than a list of facts or practices, but that it is, in fact, expressed uniquely in each individual (Moran, 2001).

Moreover, Skype interviews provided advantages for these teachers with respect to conveying students a message about the importance of English language learning as it is internationally-used. The students had a chance to use English for an authentic purpose, interacted with somebody in English in real time, received authentic feedback and were engaged in authentic communication. In addition, they realized that people speak English internationally, and people from different cultures have their own 'English accents':

> They speak some words in Maria's interview which they told me about later on. 'She said this word this way!', and they were happy that they could figure it out, and they could understand it though it seemed a bit different than they used to hear it from me maybe or than they thought it might be. They could somehow see how English is, it's all English, but how it's different, my English is different from Maria's and a native speaker's and so on. (Ayat)

Maria also explained that it was very motivating for students to listen to a teacher with a different accent in English. She noted that students were surprised at their own ability to understanding a person speaking English with an unfamiliar accent. By interacting with a non-native speaker through Skype, students realized that they were able to communicate with somebody whose native language is not English through using English.

New Cultural Understandings and Breaking Down Stereotypes

When interviewed, both teachers explained that some cultures are more "closed" to other countries (e.g., Egypt) and, as a result, it can be more difficult to have access to them. Although the Argentine students knew something about "Big C" Egyptian culture from their lessons at school (e.g., history, pyramids), this project provided them with an opportunity to interact with a cultural representative and learn more about the everyday culture, or the "small c" culture of Egypt. Ayat felt very pleased to be a part of this important mind-opening experience for Maria's students, and she felt as though she assumed a role of a "cultural ambassador" of Egypt for Maria's students.

> That was very important for me too, yes, because as Maria said, people usually just think about when it comes to Egypt, what comes to their mind, the stereotypes of camel, desert, and so on, but Egypt is modern. Egypt is completely different. So these kinds of Skype interviews are very important for me to introduce how modern Egypt is, and how life is just like any other place. How it is different than these kinds of stereotypes people get from maybe books, media, and so on. So it's very important, as it was very important for Maria to change these kinds of stereotypes in her children to be, as she said, more open-minded and more understanding of how real things are. (Ayat)

It was not only the students but also the teachers who gained new cultural understandings about the "small c" cultures of the other culture through these contacts. Both teachers explained that, although they had general information about the other culture, it was through these intercultural exchanges in their interviews and their personal chats that Ayat learned that "every time I talk to her she says it is a festival, it is a certain day; and [the weather] is exactly the opposite of Egypt; when it is very very hot here, it is very very cold there" and Maria learned that "traffic [in Egypt] .. is disorganized, and [Egyptians] eat koshary [Egypt's most famous dish] everywhere!" When they saw Ayat for the first time on Skype, many of Maria's students were surprised by her appearance, and they asked Maria: "*Why is she covering her head?*" Although Maria had provided her students with background information about Ayat and their friendship prior to their interview, it had not occurred to her to inform the students about her religion or her appearance because she did not perceive Ayat as different: "I never expected their reaction, because I saw Ayat every day, I saw her pictures everyday, so for me it was normal" (Maria). Later, Maria also explained how this project helped students to develop their own perspectives about the cultures, and be critical about stereotypes:

> I have one student that goes to a [Catholic] school, she said that the nuns talk about Muslims. She said when we interviewed Ayat, she talked about her religion and this and that, it's like the nuns told them different things. So for me, it was good that my student remembered Ayat's interview because now they can compare and they can talk about Ayat's life. So, that's why, for my students, it was a mind-opening experience. (Maria)

Maria's student had prior ideas about Muslims (ideas that were transmitted from her schooling) but because of her direct experience of speaking with Ayat, the student learned to question – rather than to automatically accept – information provided by others who may be cultural "outsiders," and who may not provide the most accurate or complete information about a different cultural group.

Continuing Culture Exchange via Facebook

Pleased with the success of their reciprocal Skype interview exchanges, each teacher made a digital video with images on PhotoPeach (<http://photopeach.com/album/qgkfv4>) of the project and shared it with the larger WiA community. The Figure 1 below is a screenshot taken from the video to illustrate how students interviewed each teacher.

Figure 1: Maria's students, as they interview Ayat.

Fascinated by their two recent members already telecollaborating, Webheads provided further encouragement through their comments both in the email list and within the Photopeach video in response to Ayat and Maria's telecollaborative classroom cultural exchange. Moreover, Ayat and Maria's own students developed a curiosity and interest in interacting with the other teacher's students, and asked to have student-to-student interviews. However, because of time zone difference, this was not possible to implement during school time. Therefore, with the encouragement from their WiA friends and their own students, Maria and Ayat decided to expand the project. Tapping into their students' existing social media practices (i.e., "Since our students spend most of their time on Facebook playing or sharing some stuff," Ayat), the two teachers created a Facebook group, called "ArgentEgypt." The group consisted of 67 members including the two teachers and their students.

Given that students had already gained a reference point with the other culture through their interviews with the teachers, their cross-cultural exchanges could continue – both between teachers and students, and among students across both groups – as Ayat explains:

> Then we started talking with our students about many topics. We had some talks about… festivals in our countries and some traditions and we shared it together. [….] And it worked. Step by step we started encouraging them to go on and write and share. They liked the group so much and they made friends. My students made friends with Maria's students, and they started chatting. Sometimes we go in the group, me or Maria, and start chatting with the students. We talk about a range of topics, food, traditions, festivals, things to those cultures, some language points, discussing for example, are you an early-bird or a night-owl and what does this mean and so on. And they respond to it. (Ayat)

The Facebook page provided students another opportunity for interaction, in which they could practice realizing important social functions of language (such as expressing greetings, extending birthday wishes, or learning abbreviations commonly used in internet communication) as well as learning about each other's customs and traditions. Teachers also joined the interaction by posing questions related to students' lives and interests (*"What's your favorite sport/game/color/food?" "What do you want to be when you grow up?" "What are your summer plans?")*, and provided information about national holidays and national anthems.

All these cross-cultural interactions through Skype and Facebook provided students and teachers to interact in real communication with cultural insiders, which is an important addition in language learning classes. Rather than discussing culture and their interests with their classmates through textbook prompted questions for the sake of language practice only, the students were able to use English for real purposes in each step of this collaboration.

Sharing Language Practices

Allowing students to use English for real purposes and communication with real people from other cultures, (i.e. cultural insiders) was the main objective for Ayat and Maria when they started their multiple-step telecollaboration. That is why, when they created ArgentEgypt Facebook group as an out-of-class experimental project for further real English practice and intercultural exchange, they did not know how the project would unfold. Likewise, throughout their interactions, the students' home languages also emerged as a topic for discussion on Facebook. Although students were required to interact in English on the Facebook page to ensure equal understanding by both groups (i.e., and not in their native languages among themselves), as they became increasingly interested in each other's cultures, they also began sharing a few words in Arabic, or Spanish. It was through these interactions they learned that their home languages even share some words in common, such as "*batata*" (sweet potato). Other exchanges among students on Facebook also illustrated how students attempted to use the other group's home language. For example, one Egyptian student wrote a comment under a picture of one of the historic places in Argentina that said "*Viva Argentina*"; and another Egyptian student posted an image of a piece of paper with the expression "*te quiero mucho*" (*I love you*). This introduction of vocabulary items from the students' native languages functioned as the sharing of cultural resources – thereby broadening students' collective linguistic repertoires. Students' use of other groups' native language had an inclusive function.

However, because English is the only shared language among students from Argentina and Egypt, most of the posts needed to be written in English. Therefore, in addition to a meaningful intercultural dialogue with age-peers from another culture, Ayat and Maria simultaneously created the conditions for authentic, meaningful practice with the target language, which helped students to develop new understandings about language and language learning, and not to perceive learning a language as a formal subject, but learning the language as tool for global communication. In addition, both teachers believed that the telecollaboration gave their learners confidence and encouragement to continue learning and using English. The excerpt below, taken from the group interview with both Ayat and Maria, highlight these dimensions of authentic, meaningful language practice, as well as students' raised awareness of the utility of English in global and intercultural communication.

> Maria: As Ayat said, it gives them confidence *to go out to the world.* [...] I guess they realized that they know English.
>
> Ayat: Yes, right. And that *their English is real. It can be used, and it can be understood by other people.*
>
> Maria: I guess that they reassured what they were doing in classes: "Wow, I can do it!". Because when I said "Can you speak English?" [asking them to use English during class time], they would say "No, we can't." So *when they speak, you say, "You see, you can!" That makes the difference.*

In sum, both educators were invested in this telecollaborative project because it enabled students to interact with one another, not only to practice their English, but also to go beyond the classrooms to learn about other cultures: including customs and traditions, as well as more quotidian social practices. In other words, both teachers believed that this telecollaboration provided their students with a meaningful learning experience, and one which fostered a sense of open exchange and curiosity about one another's cultures, which is an essential step towards social justice in a globalized world.

Discussion

In today's globalizing world, culture is a resource to address and solve social problems such as conflicts between societies, communities, etc., and thus, it serves as a powerful resource to bring about social justice (Sorrells, 2013). In that sense, intercultural exchanges through telecollaboration can play an important role in the promotion of social justice. In this chapter, we have illustrated how telecollaboration facilitated a global collaboration among educators (members of an online CoP), as well as how it facilitated an intercultural dialogue among two teachers and their classes. We believe that these have important implications for social justice and here we discuss what it is about this particular community that makes it possible.

WiA is an organic community in that there are no clear-cut role definitions or expertise divided among its members. This enables each member to occupy an expert position throughout their development, which is something that is usually more restricted in traditional institutions of education as well as in online communities with "top-down" organizational structures, where technical experts and leaders of the community are pre-designated. Perhaps the most important characteristic of WiA is that membership in WiA is free, and open to all English language teachers anywhere in the world. In contrast, it is more difficult to talk about true equality with for-profit online workshops and courses, where individuals must pay in order to participate. Furthermore, the voluntary work carried out in all activities of this community ensures that true equality and social justice emerge as core characteristics of the community itself. Also, the model of voluntary and everyone-is-an-expert participation in this community allows each person to grow within the community; all are expected to learn, grow, and participate. This enables participants to "own" the activities and values of this community, and to recognize everybody's expertise on equal terms. Therefore, it is a model of participation that levels the field: rather than a "join and follow the expert" model, WiA can be characterized as having a "become a participating member, become a leader" organizational structure. Webheads value each other's experiences and knowledge, and encourage each other, even the most novice members to share their work with the community – as was illustrated with the collaboration that took place between these two teachers. Moreover, this emphasis on participation of all members encourages a sense of "giving back" to the community. Members feel that their professional development reaches a higher level as a result of the help of this community. In turn, they themselves begin to take on more active roles in the community's organization and leadership when such an opportunity presents itself (e.g. moderating BaW workshops). All of these characteristics demonstrate how members in WiA are treated fairly and on equal terms as long as they are willing to contribute and share. Any ESL/EFL teacher, or anybody interested in this field, is welcome to join this community, regardless of their race, gender, ethnicity, religion, or language. These principles are produced and reproduced by members of WiA, and it is likely that they even find their way into members' own pedagogical practices.

Social justice also means having equitable access to and distribution of resources and services. WiA, in that sense, embodies these principles, by offering all of their activities and resources through open source tools. Thus, if one has the necessary tools and access to join this online community, presumably it means to have the access and opportunity to use all of the free tools that everyone else has access to. This, in turn, encourages members to become advocates of using open source technology tools with their own students. For example, in the case we described above, these teachers had limited technology resources at their institutions. Using open source technology tools in these projects enabled them and their students to communicate, exchange cultural information with, and indeed, to learn from, people on the other side of the world for free.

More specifically, our case study example of two teachers shows how the international scope of this communication and collaboration helps to "break down barriers" such as stereotypes and pre-existing beliefs about various cultures around the world among individuals from different cultural backgrounds. We believe that this serves as a key foundation for social justice among the groups involved. In this sense, the social justice principles and practices embedded in the WiA community are extended to participating teachers' own teaching contexts, and then to the interactions among students of these teachers.

Kramsch (2004) characterizes a language teacher as a "go-between among people with various languages [...] as a cross-cultural mediator, someone who has acquired the ability to interact with others [...] who has learned to accept other perspectives and perceptions of the world, to mediate between different perspectives" (p. 44). In our case here, these two teachers provide powerful examples of this mediating potential of language educators, by collaborating with each other through a language that neither speaks as a native language. What made this possible for the two teachers (who had never met in person at the time of the group interview), is the online community that they belong to, and the technologies they used to communicate and collaborate with each other. Accepting each other with their differences and learning about each other's culture, they provided a model – as well as an opportunity – for their young students in both countries to become aware of the similarities as well as the differences that make us all human.

Finally, this telecollaboration had empowering effects for both the students and teachers involved. Teachers' and students' interest in learning common expressions in the language of the other group positioned these English language learners as "multicompetent language users" (Cook, 1992), as opposed to merely language learners with limited English proficiency. Particularly for the 10–11-year-old students, this may have been their first contact with others who were interested in learning more about their own culture and language, thereby constructing them as "knowers" (rather than only learners). This project enabled all participants involved to experience their own multilingual repertoire and to feel ownership of their own languages. As Vetter (2012) argues, this also gives language teachers an opportunity to further affirm the rich linguistic resources that students bring to the foreign language learning process, and thus use these resources to cultivate students' multilingual identities. Moreover, these teacher-to-students intercultural interviews also empower non-native speaking language teachers. Through these exchanges, students gain a first-hand experience of their teachers as legitimate English speakers. Experiences such as these may also generate an understanding of the fact that, in our globalizing world, English does not only belong to native speakers (McKay & Bockhorst-Heng, 2008; Widdowson, 1994).

Coda

Ayat and Maria's intercultural and telecollaborative story does not end here. Since their first contact in January 2011, they have remained in frequent contact with one another, not only seeking ways of collaborating with each other but also building an exemplary transnational friendship, which opened up new worlds to them as well as to their students. As this chapter was being written, Maria had already bought her ticket for a 10-day trip to Egypt in December 2012 to finally meet with Ayat in person – after 2 years of ongoing computer-mediated contact. During her visit, Ayat plans to take Maria to her school and to gather all her students who interviewed Maria, so that they too, can meet her in person. Although these students have already graduated, they will return to their previous school, meet with Maria, and have an opportunity to realize how small the world really is.

References

boyd, d. m., & Ellison, N. B. (2007) Social network sites: Definition, history, and scholarship. *Journal of Computer-Mediated Communication*, 13(1): 210–230.

Cook, V. (1992) Evidence for multicompetence. *Language Learning,* 42: 557–591.

Kozinets, R. (2010) *Netnography: Doing ethnographic research online.* London: SAGE.

Kramsch, C. (2004) The language teacher as go-between. *Utbildning & Demokrati,* 13(3): 37–60.

Kulavuz-Onal, D. (2013) *English language teachers' learning to teach with technology through participation in an online community of practice: A netnography of Webheads in Action.*(Unpublished doctoral dissertation). University of South Florida, Tampa, FL.

Liu, H. (2007) Social network profiles as taste performances. *Journal of Computer-Mediated Communication, 13*(1): 252–275.

Liu, H., Maes, P., & Davenport, G. (2006) Unraveling the taste fabric of social networks. *International Journal on Semantic Web and Information Systems, 2* (1): 42–71.

McKay, S., & Bockhorst-Heng W. (2008) *International English in its sociolinguistic contexts: Towards a socially sensitive pedagogy*. London: Routledge.

Moran, P. R. (2001) *Teaching culture: Perspectives in practice*. Boston: Heinle & Heinle.

Sorrells, K. (2013) *Intercultural communication: Globalization and social justice.* Thousand Oaks, CA: SAGE.

Stevens, V. (2007) Issue: Tools for online teacher communities of practice. In: Hanson-Smith E. and Rilling S. (eds.), *Learning Languages through Technology*, pp. 257–269. Virginia: TESOL.

Vetter, E. (2012) Mulitilingualism pedagogy: Building bridges between languages. In: J. Huttner, B. Mehlmauer-Larcher, S. Reichl & B. Schiftner (eds.), *Theory and practice in EFL teacher education.* Bristol, pp. 228–246. Britol, UK: Multilingual Matters.

Widdowson, H. G. (1994) The ownership of English. *TESOL Quarterly, 28*(2): 377–389.

Dominique Vinet

Online collaboration with the language-learning software @genda 2.0

Introduction

From active knowledge creation to social networking, Web 2.0 technologies have widened the horizons for L2 teaching and learning. Blogs have already been largely used as asynchronous modes of computer-mediated communication and acclaimed as tools promoting the learners' autonomy and allowing them to reflect critically upon their content (Richardson, 2005). Dynamic and interactrive networks have been created, allowing users to produce and update content, and researchers working on interactive teaching involving students of different nationalities showed that group blogs "created a social workspace where students worked collaboratively to share knowledge and exchange ideas" (Lee, 2010).

The objective of teachers using ICT is not only to enhance writing fluency and literacy skills but also to gain cross-cultural awareness. Yet the confrontation of ideas may remain rather superficial if it is limited to writing, even though it gives users time to think and to produce error-free content. At the other end of the spectrum of ICT tools, recently developed chatrooms now allow users to exchange ideas orally and synchronously, making the socializing process more difficult as immediate responses are required. Chatrooms may also prove more intrusive and generate unwanted affects due to eye-contact and the non-verbal communication skills used by the interlocutors, thus posing the problem of their adequacy as a socializing tool.

The objective of this chapter is to examine a collaborative project on comparative law involving French and American students from Bordeaux and Chicago, and show how we carefully phased the use of a variety of Web 2.0 tools with a view to gradually developing cross-cultural awareness while enhancing their language skills.

Inspired by Montesquieu, the founding father of democracy, we decided to test the ability of French students and English-speaking students in law from Illinois to question the way democratic concepts have been shaped into statutes which, in a number of cases, are so much at variance. Criticizing differing views requires knowledge and talent but needs no personal interchange. Discovering the cultural and historic reasons at the origin of the decisions made by legislators requires insight into values that the global village has buried under a layer of uniformity, and wilful collaboration on both sides. To foster intercultural exchanges and educate these young people for world citizenship, it was necessary to bring them together, but for practical reasons, exchanging ideas across the borders required the use of online tools and a careful process leading from observation to tolerance and mutual understanding. Although collaboration between the universities of Bordeaux and Northern Illinois had been going on for more than fifteen years under the form of summer schools on the basis of a student exchange, the opportunity had always been limited to those who could afford the expense, no grant being offered to participants. Part of the motivation in conducting the experience was therefore to offer low- to middle-income students the opportunity to share ideas with American students with a different social background.

A Double Target: Developing Language Skills and Understanding Cultural Differences

The academic objective of the project was to get the students to exchange ideas and produce reports on legal subjects involving comparative law. The experiment was held at master's degree level but could have been conducted at undergraduate level. The idea was to get them to learn from each other and teach each other, and gradually become aware of the divergent or concurrent approaches under the supervision of teachers who were able to follow up on the work produced, correct the vocabulary lists constituted, the notes written and the reports published, whether they be in written, oral or video form. The experiment was carried out from January to April 2012, with final reports to be delivered in June.

The original idea of the project stemmed from the request by a colleague from the University of Northern Illinois who wished to bring his law students into contact with French students during a summer program held in Bordeaux in June, in which they were to be taught European law, and an introduction to French constitutional law. It was therefore decided that the French students' reading for a master's degree in public law would be included in the project, and students would be expected to hand in reports on subjects requiring a comparative approach of the law. The project was set up on the American side as an option in which students would earn three credits. An agrement was reached on six controversial subjects.

French students in law are expected to learn a foreign language – mostly English – to participate in international conferences during their doctoral studies and by so doing, widen their horizons. The aim of the project was both to develop the language skills of the French students and link their mostly technical knowledge of legal English to a better understanding of the cultural, social and historic background presiding over the choices made by foreign legislators and the formal applications of the law through procedures, legal documents and legal practice, with the hope of generating distantiation and mutual understanding of the intent of the law on both sides. The outcome of the collaborative work produced was not sheer simulation and could therefore not pervert the action and the use of the language as is the case when communication is simulated (Ollivier & Puren 2011).

One of the major problems that French students usually discover when they examine the American legal system is the importance of procedure and form in general, and the comparatively little space left for the social, cultural and family background of the offenders, as Jock Young explains in the theory that he calls Left Realism Criminology (Young 1987). Conversely, public law is all about protecting individual rights against the intrusion of collective interest.

We therefore agreed on six different topics on which the American and the French have widely differing views and legislations, and which were to be studied in pairs:

- Money in Politics: Financing National Elections,
- Separation of Powers
- Judicial Independence
- Guns and Individual Rights
- Separation of Church and State
- Role of the Government in Health Care

The underlying theme stood out as a democratic challenge: how can the law deal with antagonistic interests, as well as offer a social model acceptable outside the boundaries of the country concerned? Beyond the legal approach, moral values were at stake and the relationships of religion to money and politics interfered with the principles of democracy and social justice, notions which required clarification and a distanced view enlightened by the history and culture of the two countries.

Didactics has shown that the best way to learn languages is in context, through practice and repeated contact with notions and vocabulary that will help with what Husserl called the sedimentation of knowledge (Husserl 1938–1962). Students were therefore asked to share and mutually correct their written assignments and oral recordings throughout the development of their competences. The process was expected to arouse their awareness of differences in approaches through the confrontation of viewpoints and debates on points of contention. To avoid academic exchanges, which might have been appealing to teachers but could prove boring to the students, it was decided that the project could take different forms including role plays in which the French students would explain to their classmates the viewpoint of American students supposed to defend the American choices in controversies staged as mini videoconferences. Three questions were then raised: how could debates be organized online, what software could be used to prompt both oral and written exchanges, and how could the work of the students be monitored and assessed during the process?

The last two questions could be solved easily. We had developed @genda 2.0, a piece of software meant for teachers of languages to create their own courses for either contact or distance teaching (the interface comes in three languages – English, French and Spanish), and for students studying languages to learn and practice online. Students on both sides only needed to register at <http://www.agenda-2–0.fr>

and key in an agenda code and a group code that was given to them by their teachers. Both the American and French students were enrolled in the same group.

@genda 2.0: an Online Collaborative Platform

The software is an e-learning platform which can be fed by teachers, and the students can work on it either individually, or collectively, in pairs or as a group. It was originally developed to allow students unable to attend courses for medical reasons, or with no financial support and forced to take a full-time job to meet the tuition fees and other costs; it is now being developed under a European programme for the teaching of French as a foreign language in North-African countries.

After the initial registration, users can directly log on to their workspace. The screen is divided into three windows: The main window displays the assignments for the different work sequences, and links to the different documents provided or self-assessing exercises, written and/or oral. Clicking on links displays the documents, and the user simply needs to click on the red button indicating the active sequence on a narrow column of twenty-four buttons on the right to get back to the original window. It is thus easy to switch from the assignments to the required documents and back.

On the right is a tool-box including access to the web for information, several dictionaries, lexicons, a voice-recorder, text-to-speech modules – to be used as a prompt to help users better pronounce whatever they wish to say in the foreign language used –, an applet allowing them to send video mails, and a forum where the users can exchange ideas collectively, ask for advice and help, and seek cooperation. The users also have access to a chatroom implemented in the software and can share an online interactive whiteboard showing instantly any modification of the document shared. The students simply need to agree on what time they log on together.

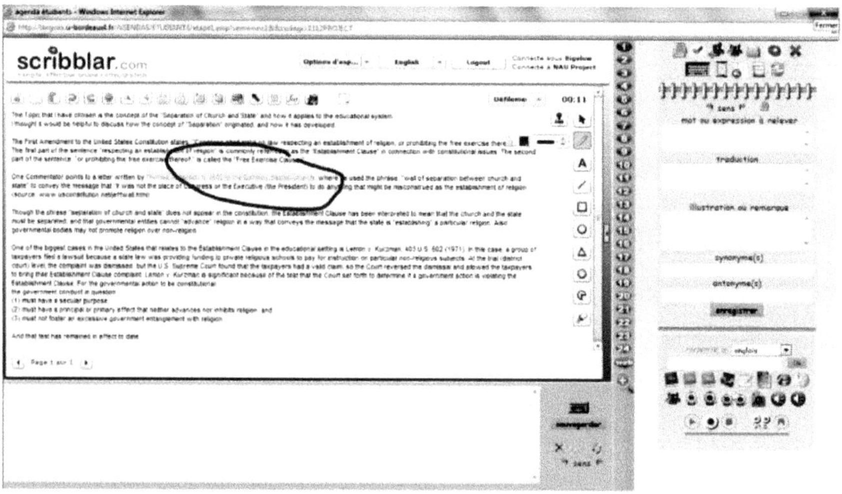

Figure 1: Screenshot showing interactive use of the whiteboard.

The tool-box also provides a module to pick up vocabulary which the French students are supposed to contextualize – provide sample phrases, translations, synonyms and antonyms – and which can be revised and enriched or corrected at the click of a button. Students may also revise and enrich their vocabulary list, have access to reference documents or grammars on line, get direct access to their classmates' notes, and they may also reserve subjects for their reports if needs be.

The bottom window is a notepad where they are expected to jot down notes, work out questions and answers, that they may rewrite at any time. These notes can be read and/or modified or corrected by the foreign students to whom they have given a password allowing them to access their notepad. They only need to ask someone to collaborate, either by individual or collective e-mails – standard ones or video mails –, or even through the ooVoo chatroom. They have a list of all the students involved, showing their personal details and their photos, and they may collaborate with one or more, and even change partners at will. After the first draft, they may wish to collaborate in the writing of a fair copy and they can choose to use another device, "Google docs", which is also accessible from the @genda.

It is certainly harder to spur oral than written exchanges as they require a capacity for debating without taking time for reflection; yet, oral expression is just as basic to the learning of languages as it is to debating, and the objective of the project being also to develop the understanding of cultural differences, direct confrontation of ideas on the chatroom was encouraged and organized. Considering the time difference between Chicago and Bordeaux, slots were determined for connections. They were necessarily private, but the teachers could listen to and supervise the exchanges and even call on the students to start one under their supervision, at their request, which was not done, to preserve the privacy ot the students. At any rate, online conversations could be recorded with the voice-recorder embedded in the software, and stored in the student's file-box or even in the appended online language portfolio, where reports were to be stored once they had been completed in June. As an alternative, Video mails were also sent to correspondents, allowing asynchronous oral exchanges. The URL address of the video mail could be copied with a right-click on the link provided and pasted in the @agenda 2.0 either directly or as a reference document. The senders were warned that their videos could be used collectively in class or individually by other users of the software. The project developed through a fourfold approach: bringing students together, suggesting controversial topics to spur intercultural exchanges, helping them develop work habits, and providing a feedback.

Phasing of the project

Careful phasing was necessary at this early stage since any mishap could easily dishearten the participants. Teachers on both sides therefore needed to make sure that registration had been completed on time and that all data had been made available. The French students were given a week to register, key in their personal details, and open their online portfolio of languages. The specificity of the portfolio is that apart from the classic features offered by other online portfolios, it allows teachers to validate the levels of competences declared, and includes a dossier where documents of all sorts can be stored and opened by any visitor allowed to log on it as evidence of the owner's competences.

Students were requested to rapidly select a topic for their project, but they were left free to change in the course of the term if the topic ended up being of little interest to them, which did not happen. The six topics agreed on were proposed and no more than two students were allowed to select the same. Additional topics could be added at the American or French students' request but none of the participants seized the opportunity and they started developing such themes as the differing ways in which individual rights are protected in France and the USA, or why bearing arms being historically considered as a privilege of the nobility in France, has never been considered as an individual right.

The group of French students being already constituted and working in weekly classes, a one-hour session was enough to explain how the software worked and what the students were expected to achieve. The American students were given assignments during a tutorial organized by their teacher. The French students were to take initiatives but not unexpectedly the American students took the lead by selecting the subjects proposed and using the blog to seek partners. The American students immediately found motivation in the pleasure and interest they would derive from the interchange. It is important to note that the American students involved were not seeking to develop their language skills but to extend their understanding of French law through a comparative approach. As they were communicating in their own language, they did not feel any reluctance and proved keen on helping the French students improve their language skills, but they asked for clarification when conceptualization led to intricacy or hesitations.

Experiments have shown that online communication differed from face-to-face communication, with shortened statements and spells of silence between them (Marquet & Nissen 2003), and the French students had to learn how to express themselves clearly without stammering. They needed to make sure they would be understood and made use of the tex-to-speech modules provided in the @agenda 2.0 to improve their pronunciation.

Figure 2: Sitepal offers good-quality text-to-speech facilities with a variety of accents. <http://www.sitepal.com/> Sitepal demo module is used in @agenda 2.0. The limitation is that no more than 200 characters can be inserted, but this is enough for students to practise.

During the second class, each student explained orally which subject they had selected and they were asked to give their own view. The next step for them was to make contact with their American counterparts. At that time, the American students had registered and their personal details were available on the list of online contacts. The next move consisted in selecting one or two mates to collaborate with. Contacts could be made by direct or collective e-mails, or through the blog. Teachers on both sides made sure that the topics were evenly distributed and that no one had been left without a partner.

In parallel with the work done in class, 20–30 minutes were then devoted every week to the development of the project throughout the term. Depending on the time and day of the class, ooVoo exchanges were to be arranged during classes but the only vacant slots being in the morning in France, no videoconference could be organized in class and all of them had to be made on an individual basis.

In a second stage, the French students were expected to share their views with their American friends, either by giving them access to their notes and the papers they had written, or by using the interactive whiteboard provided, or even by exchanging ideas on the forum. Each of the three media offer different features:

– Notes are personal and can be shared with selected persons. They may contain errors which can be corrected privately, and students who felt shy about publicizing their views or letting others know about their linguistic deficiencies were obviously more comfortable with private exchanges.

– The forum is meant for short messages requiring quick responses and was supposed to be mostly limited to making contact or asking for help, yet most of the exchanges were originally made on the forum by the Americans who sometimes posted very long messages. They had to be redirected to the other tools provided.

– The whiteboard, and the file-box allow users to propose documents for the group after validation by a teacher, who can check them more conveniently than when Google + tools are used.

Sharing views being the first stage in the collaboration process was a delicate moment which could determine the interest felt by the students and had to be carefully staged. Differing views could lead to the ossification of positions, cutting debates short. What mattered at that stage was to observe differences in the French and American law and procedures, and to determine what the object of discussion would be and lay the corner-stones of cultural interchange.

For efficient cultural interchange to take place several ingredients are necessary:

– A good knowledge of the system and values of the foreign culture, more specifically of the American legal system;

– Emotional experience toward the foreign culture and the people who acknowledge it (Andreeva-Sussin 2009);

– Know-how, a capacity to question one's cultural references, a sense of nuance, and the ability to express one's feelings and seek similarities and differences between interlocutors so that transcultural constructs be finally achieved.

At first, considering the cultural gap to be bridged, students were tempted to write down their ideas rather than exchange them orally, which required a clear view of the subject and a specific vocabulary that they did not master yet. The notepad, coupled with the e-mail box, appeared as major tools in the cooperation process since notes could be shared and they were supposed to show the student's reflection in progress. This is where each of them finally confronted their ideas with their foreign counterparts in writing, and where opinions, tentative solutions and gradual concurrence showed up. It allowed students to develop more personal relationships with a limited number of partners, contrary to the forum where everyone could have a say and exchanges happened to be extensive but probably less focused. Yet, the content of the notepad remained work in the making, a first draft, to be shared with a limited number of partners. Once the notes, or part of them, were gradually worked out into a presentable document, students – the Americans first – proposed them as reference documents. A button in the tool-box opens their file-box and clicking on an icon allows them to propose their documents to the group. As collaboration tended to develop on a one-to-one basis, an online interactive whiteboard was then introduced.

The whiteboard is where the whole group, or only some students, may share collective documents to be enriched and modified at will by all invitees, the original author keeping a high hand on it and allowing select partners to collaborate, without teachers eavesdropping. The fundamental concept is that students should be given enough leeway to develop arguments of their own without fearing the censure of teachers, yet enjoy their help in two essential phases of the development of their work: the starting point, and the final delivery. Google docs were first considered as the best option to create collective constructions open to a selection of students. They seemed preferable to standard wikis since they are both less impersonal, although entries or corrections are anonymous, and they do not expose the writers to the criticisms of the whole group, so the interpersonal link involved in sharing the notes is not lost to the free-wheeling anonymity of online commentators but it soon appeared that the students were not familiar with using them and expected to be able to work synchronously as if they were present in the same room, which the whiteboard allowed.

The Notion of Collaboration and Connectivism

Before expatiating any further on the construction of meaning, we need to ponder on collaboration as distinct from cooperation. Collaboration differs from cooperation in that the creative acts performed require the emergence of shared understandings developing jointly. Conversely, individuals involved in cooperation needn't engage in a joint negotiation – from which shared understandings emerge – they may simply execute instructions willingly. The process has been analyzed by connectivism, a recent theory of learning introduced by G. Siemens and S. Downes. One aspect of connectivism is the use of a network with nodes and connections as a central metaphor for learning. They claim that philosophers have never agreed on a clear definition of knowledge, which is supposed to be the ultimate aim of learning, for it stems from "the complex interplay of our knowledge definition and our world views (religious, social, political, and so on)" (<http://www.connectivism.ca/>).

L. Johnson *et al.* (2010) consider that the development of networking – or should we call it net-working – contributes to the co-building of knowledge through connectivism, and should lead to more cross-curricular approaches. Our project aimed at offering an opportunity to develop language skills and legal knowledge while allowing the students to take a critical stance, which means that they needed to learn to question the authority of texts and use their own ideas to construct meaning, relying on their own moral values and the concept of social justice that they had constructed both intellectually through the courses in public law they had attended and politically through their personal commitments.

When it comes to legal documents, the difficulty to admit that the law conveys no sacrosanct universal truth is definitely hard to admit unless a solid rationale is brought forth to feed meaning, but the challenge extends to trying to fit notions into words and using words to discuss principles or processes, or even functions that have no correspondence in the foreign country. In the detailed list of topics proposed to the students, some words and phrases were left in French, like "*laïcité*", which has no accurate translation – secularism, secularity, religious neutrality – and cannot be understood fully without careful definition and exemplification. In most cases, understanding will develop from interchanges, but theories dealing with socio-cognitivism insist that some form of monitoring must be applied by teachers. Therefore, the place of the teacher in an online collaboration process remains the key to success.

The Teacher as an Online Mediator

Research has, for a long time, placed more emphasis on fostering a greater sense of awareness of the various strategies used to produce the final product (Pennington & Brock, 1992). But what was technically feasible in the nineties was far from allowing any insight of the teacher into the collaboration process. This is certainly one of the strong points of the @genda 2.0. The software enables teachers to get closer to the – sometimes slow – elaboration of their students' thought as they can have a look at the work in process on their notepads or on the whiteboard. They may even spur the process through questions or links to further information that can be included either directly in the notepad, or sent by e-mail, or even written in the notes of appreciation available. They can also suggest questions to be asked to the partners and gradually develop the students' awareness and critical sense through maieutics that would be both applied to and by the students. Yet they must make sure not to interfere too directly in the debate, for that could deter the students from going any further.

Just as the teacher can read and if necessary correct the notes in the notepads, he or she can keep an eye on the vocabulary list which will be put together by the students as they discover new concepts. There is a chance that some of them may write things wrong and that it will hamper the progress of their reflection. It is thus most important to make sure that they do not misconceive anything and that the ill-understood or ill-written vocabulary which will still be active in their working memory will not be ingrained in their long-term memory when they move on to the next stage, the oral presentation, or the drafting of the final report. The role of the teacher as a mediator highlighted by Narcy-Combes (2005) should therefore not be limited to giving assignments and prompting students to communicate and exchange ideas; teachers should be fully involved in the learning process which may be widely overlooked when students have overcome the inhibition present at an early stage of their interchanges.

The working out of material for task-oriented teaching is often considered as some form of mediating or propping (Jordan 2004); by interfering at the initial stage of the learning process, the teacher anticipates the learners' reactions and provides a frame which will set the efficient processes on the right track at the appropriate time. This will keep the individualization of learning and the distance separating the learners and the teacher from generating disinterest that could not be compensated for by occasional reunion.

In the process described above, personal contact is maintained and the interference of the teacher is assumed as a help rather than a transmission of knowledge or the assessment of a formatted piece of work. The greatest difficulty with the French students involved lied in the paradox of law-teaching at the university. Just as the French legal system is supposed to put the emphasis on legal principles stemming from democratic ethics as distinct from moral values, teaching the law assumes that form prevails over substance, or at least that form is binding. Law students are therefore seldom prepared to give leeway to their critical spirit or even instill some suppleness in their approach of the law. Our experience shows that French law students tend to stick to set forms of written expression, and lack the proactive attitude of their American counterparts who have been taught to participate in debates at a very early age and do not hesitate to question their teachers during lectures. It was thus necessary to limit the interventions of the teacher who was to operate mostly as a counselor (cf. Curan 1972), not only in linguistic matters but also in socializing.

Getting the French students to communicate both orally and in writing served this purpose insofar as the American students were supposed to correct their mistakes. The deal was made clear from the beginning: although work in the making is bound to contain mistakes, which could be corrected online by the teacher, provided the document was recorded in the notepad or on the whiteboard, every document produced and every oral presentation was to be flawless. Yet, to avoid the delusion of perfection provoked by complete correction by peers, teachers need to detect errors and to lead the students writing in a foreign language in their efforts to improve their language skills. Grammar mistakes were dealt with during the written preparation, since it would have been unwise to correct them during oral interchanges. That is why the first stage during which the students prepared questions to be asked to the Americans was crucial. Mistakes necessarily arose and were corrected before they became ossified in the memory of the learners. We could then either individually explain what was wrong by writing advice in the teacher's pad, or direct the students to one of the online grammar lessons and exercises provided in the tool-box of the @genda 2.0. When vocabulary had been misinterpreted, which showed in the vocabulary list, we would correct the translation and/or notes and provide examples to illustrate the difficulty. Alternatively, recurring morphosyntactic problems could be briefly dealt with during tutorials.

Action research has shown that "students did not often re-read their productions and hardly ever listened again to their oral ones" (Brudermann 2010). That is why students were asked to draft questions in advance and any new development in their reflection was to be written down or recorded in video or audio form, to allow the teacher to follow up on their work and promote qualitative progression in their L2 interlanguage. The principle lay in the teacher's helping them proceed to self-correction and mutual correction in collaboration with their American partners. Yet this could only be achieved if the students had become more confident with both written and oral tasks before they actually reached the second phase of the project during which they were expected to make direct or semi-direct oral contact with their partners. "One feels like a different person when speaking a second language and often acts very differently as well" (Guiora & Acton, 1979: 199), which is liable to trigger "existential anxiety" (Brudermann 2010). As a result, students tend to camouflage their deficiencies under intricate forms of speech which blur the message conveyed, and may discourage collaboration. They were therefore induced to shorten sentences, use linking words and accurate vocabulary that they could look up in the lexicons provided in the tool-box. They were also asked to use prompts for the oral presentations in class and the video e-mails they recorded. The prompts being written in the notepad, we could review them, unless they had been recorded too late.

In the early stage of the learning process, the use of the class blog was encouraged as it seemed to be the adequate tool to help students increase their awareness of the target audience "as they think critically about how their view might be reflected upon by the readers" (Lee 2010). Lee, referring to Wagner (2003), claims that "learners experience a sense of self-discovery and personal growth during the reflective period" and concludes that they increase their motivation to write because they do so for a broad audience. This is true to a large extent, but we must keep in mind that writing in a blog preserves some form of anonymity when the members do not know each other but by name, and is far less stressful than speaking in front of a real audience for an oral presentation. Writing an entry in a blog also allows re-writing and self-correction and seems to be a first step to coping with the feeling of exposition aroused by direct contact either with the class or with the American partner(s).

A blog is certainly the ideal locus for brain-storming and does not require long developments; the shiest may only venture a few words, short questions, and will feel encouraged to go further by the answers they receive while enjoying the relative anonymity of the blog. Yet, it is the responsibility of the teacher to make sure that they actually communicate in that way, and a little spurring may be necessary. Moreover, those who do not feel comfortable orally may seize the opportunity to participate in a debate that they could not hold in class.

Building rapport

It would have been no use getting the French students to make contact with the American students without careful preparation of the topics to be debated, and without first getting them to know each other. We therefore used the technique commonly taught to students in business schools and they were told how to break the ice, to introduce themselves other than by simply giving their name and saying what studies they were engaged in. They had to learn to appeal to others, generate empathy and create a positive atmosphere in the closed group of the class as well as its virtual extension. Each student was therefore requested to create a video e-mail in which they would introduce and describe themselves and explain what topic they had selected for their final report and for what reasons. Each student was supposed to reserve one subject on the agenda and find a foreign partner to work with. Seven pairs were constituted in advance, the American students being left to select their French partner. The video e-mails were to be recorded in pairs and uploaded in the students' file-boxes. We had originally arranged to create a web page showing the topic and names of each pair, with links to the online videos. The documents were to be stored in the "reference documents" section of the agenda where the American students would be able to click on the links to watch them, but the students preferred one-to-one interchange and the idea was dropped.

Up to this stage, asynchronous activity had been privileged to encourage practice, improve fluency, and allow linguistic feedback from the teacher. Yet, there is no way one can correct recorded voices without asking the students to do it all over again, which would be discouraging; so students were prompted to use two tools from the tool-box: first Linguee dictionary, which is a treasure trove of quotations with their translations where they could find inspiration, and Acapela or Sitepal text-to-speech synthesis systems, to help them pronounce the right way.

Once the preparation work was completed, the American students selected the French pairs they were willing to work with. They, too, sent them video e-mails in which they introduced themselves in their turn and explained why they had selected their topic. Some of them were studied in class and used for oral comprehension exercises.

The next stage generated more difficulties. Time was up for more technical work involving specialist vocabulary, and difficult concepts which were to be made plain to foreigners who were not familiar with them. Short documents were first drafted and corrected before they were sent by e-mail to the American partners to start debates. The students were asked to explain the point they wished to debate and to formulate questions for their American partners. Replies were studied in class after an oral presentation by the students who had received answers.

Now that they had reached the point where technicalities were to be analyzed and discussed, students could be tempted to copy and paste elements, which might have led to a lack of creativity, errors and poor short-term (and long-term) memorization; but research shows that progress in morphosyntax is obvious even in copying and pasting (Narcy-Combes 2005, 178) whcn thc tcachcr closely follows up the work produced.

Now the French students needed to address their American friends, starting with a clear description of the topic they wished to debate on and the first questions they wished to ask them. In most cases, French students were not quite familiar with the situation. They had been taught to answer questions rather than ask them and our experience shows that all students tend to ask short, straightforward questions containing insufficient content to prompt rapid answers, which often end up in embarrassing silence. They were therefore taught to offer ready-made answers in their questions, such as alternatives to be selected or negative questions implying a positive answer, or even question-tags, which they are usually very unfamiliar with. Mostly, they needed to learn how to develop the skills of cultural competence including active listening and how to use open-ended questions allowing time for interaction. They also needed to learn how to express themselves in simple but accurate terms. This can usually be achieved through building questions starting with a statement, or a summary of the matter under discussion.

For instance, one French student was tempted to ask a question requiring some thinking and expressed in a rather abstract, theoretical cant: "What is the impact of the status of district attorneys on the accusation process?" We told him that a far better formulation would start with the statement of facts and a question on the direct implications, in clear, straightforward English, and he ended by saying: "District attorneys being elected, don't you think that they may feel tempted to charge defendants more severely to please their voters, especially in times when the crime rate is rising?"

In this case, the unsubtle form of arm-twisting involved in the negative question "don't you think that…?" could have been replaced with a more diplomatic formulation, such as "is it unlikely to think that…?" but the student was still biased with the typically Jacobin view that social justice is to be achieved by the state, not through the practise of electoral democracy. It is a common belief among French students that the law is the ultimate vehicle of social justice and that there is a risk in letting the mob have their say.

Socializing was at the heart of the matter and choosing the right form was essential in the interaction, which could easily stop short if the reputation of arrogance that the French often have abroad showed in the way they spoke. What we tried to develop at that stage was, beyond morphosyntax, and communication skills (*savoir-faire*), the interpersonal skills that are translated into French by *savoir-être*.

Once these skills were developed or improved, students were given more leeway and left to interchange freely in both synchronous and asynchronous ways, via e-mails, video e-mails or shared documents, which could be elaborated with the help of more than the pairs involved. They were then invited to take the final step, ooVoo video conferences or Skype conversations, when they felt ready. Skype conversations escaped the follow-up process and were left for the end of the development stage.

The last stage was the elaboration of the final five-page report, which was then to be completed with a *viva voce* at the end of the semester. At the beginning, students were asked to draft their reports in their notepad so that their work could be followed up and guidelines would be given. It was important for them to realize that the report must not be a synthesis of the interchange with their American partners. It was to show convergences and divergences, and distinct conceptions of social justice in general, and account for them as cultural differences and not only as different approaches to the law. The role of the teacher at this stage was not to make sure that the report was written in good English, but that all points of view were respected and considered receivable. Students were asked to have their reports reviewed and validated by their partners and the different reports were finally pieced together for online publication on the Intranet of the university department of languages.

Achieving Mutual Understanding Through Web-based Learning

This project was meant to develop learners' intercultural and social awareness, which implies "exploring where the learners are in terms of existing beliefs and representations concerning self and others" (Narcy-Combes 2009). They were thus supposed to question their beliefs and biases, and take into account a more "ethno-decentralized" vision of some socio-political aspects of the American law. The advantage of online teaching in this case was that students could be brought into contact across the Atlantic without ever meeting, while actually arranging asynchronous or synchronous debates. It even allowed the development of oral communication skills in the foreign language, which is so difficult to achieve online as it necessarily requires the teacher's or a referee's presence. But the objective was not strictly linguistic and social. The finished product, the report written individually, was not only assessed by the teacher; but recorded in the portfolio of languages attached to the @genda 2.0 where the students self-assessed their competences with the help of the CEFRL grid provided. The portfolio was meant to be visible online and the URL could be linked to an online application form for a job or internship, which anchored the project in real life.

Developing a collaborative project promoting intercultural dialogue between law students was made easy as, first, the law is not taught the same way, with a rather theoretical approach in France and a more case-based one in the USA, and second, the principles and procedures differ largely. This necessarily determines different attitudes toward events and situations. The students were brought to see differences and similarities that they observed, compared and analyzed. But did that help build cross-cultural bridges?

The danger inherent in online exchanges is that blunders may be committed, which may interrupt exchanges abruptly, in the absence of a mediator able to restore contact when it has been broken. The Virginia Tech massacre, still on the mind of the French students, recent publication on the growing level of student debt and the opposition met by President Obama on the health care reform of 2010 made the students feel that the USA had developed into a two-tier society generating violence and social inequality.

That is why careful phasing and preparation was to be scheduled before students were left to exchange freely. Whatever the topic, what mattered was to make sure that differences and similarities could be observed clearly and that the students admited from the start that there was not one truth. Web-based learning and targeted instructional strategies appeared necessary to achieve mutual understanding and the tools provided in the @genda 2.0 were used to address the issues of textual communication, which does not normally transmit all the features of cross-cultural intercourse, such as accents, mannerisms, styles of verbal communication (Amirault & Visser 2010) and learning patterns, as the software integrates audio and video connections for synchronous interchange. The project described here was developed over two months and every phase was closely observed and discussed. The assessment of the students was based on their competences in both oral and written expression but also on their ability to develop cross-cultural interest, which was harder to evaluate since the standards are yet to be invented; but the challenge was worthwhile and should provide material for further research. Online interchange allowed us to bring together young people who would never have met in normal circumstances and were given a chance to widen their horizons, admit that their definition of social justice was based on their own view of an ideal society, and develop common understanding of the complexity of their partners' society by going through the looking-glass.

References

Amirault, R. J. and Visser, Y.L. (2010) The Impact of E-Learning Programs on the Internationalization of the University. New York: Nova Science Publishers.

Andreeva-Sussin, I. (2009) Construire une relation avec une culture-partenaire: apprendre à se comprendre. *Les Cahiers de l'APLIUT*, 27 (2): 41–51.

Brudermann, C. (2010) From action research to the implementation of ICT pedagogical tools. *Recall,* 22 (2): 172–190.

Curan, C. (1972) *Counseling in Language Learning.* London: Pergamon.

Guioram A.Z. and Acton, W.R. (1979). Personality and language behavior: a restatement. *Language Learning,*19(1): 193–204.

Husserl, E. (1936) *L'Origine de la Géométrie*. Paris: PUF.

Johnson, L., Levine, A., Smith, R. And Stone, S. (2010) *The 2010 Horizon Report*. Austin, Texas: The New Media Consortium.

Jordon, G. (2004) *Theory Construction in Second Language Acquisition.* Amsterdam: John Benjamins.

Lee, L. (2010) Fostering reflective writing and interactive exchange through blogging in an advanced language course. *Recall* 22(2): 212–227.

Marquet, P. and Nissen, E. (2003) La distance en formation aux langues par visio-conférence : dimensions, mesures, conséquences. *Alsic* Vol. 6(2): 3–19.

Narcy-Combes, J.-P. (2005) *Didactique des Langues et TIC: vers une recherche-Action responsable*. Ophrys, Paris, France: Autoformation et Enseignement Multimédia.

— (2009) Développer la compétence interculturelle : un défi identitaire. *Les Cahiers de l'APLIUT* 28 (1): 93–104.

Pennington, M.C. and Brock, M. N. (1992) Process and product approaches to computer-assisted composition. In: Pennington M. C. and Stevens V. (eds.), *Computers in applied linguistics: An international perspective*, pp. 79–109, Clevedon, UK: Multilingual Matters.

Richardson, W. (2005) *Blogs, wikiw, podcasts, and other powerful web tools for classrooms.* Thousand Oaks, CA: Corwin Press.

Wagner, C. (2003) Put another (B)log on the wire: Publishing learning logs as weblogs. *Journal of Information Systems Education,* 14 (2): 131–132.

Rebecca Charry Roje and David S. Martins

Between "Pleasantville" and "My Way or the Highway": Promoting Productive Discussion of Social Justice in a Globally Linked Learning Environment

International affiliations between U.S. and foreign-based universities foreground the challenges and opportunities for promoting intercultural dialogue, internationalization and social justice in the 21st century. Rochester Institute of Technology (RIT) – a career-oriented, U.S.-based, technological institute – has opened global campuses in Croatia, Kosovo, and the United Arab Emirates. Within the competitive marketplace of higher education, RIT's leadership has identified international education as a way to expand the RIT brand of career-oriented educational activities to other areas of the world.

The American College of Management and Technology (ACMT) opened in Dubrovnik in 1997 as a result of thc collaboration among RIT, the local, Croatian public university, and the Croatian Ministry of Science and Technology (now the Ministry of Science, Education and Sport). The Croatian Ministry valued RIT's career-based approach, as well as its reputation, and sought a formal relationship to help rebuild economic and human resource capacity following Croatian independence and armed conflict in the early1990's. The college, now known as RIT Croatia, offers fully accredited degree programs in Information Technology and International Hospitality and Service Management at the Dubrovnik campus. A second RIT Croatia campus was opened in the capital city of Zagreb in 2011, offering degree programs in Information Technology and International Business.

While RIT Croatia has flourished, ties between students and faculty at the Dubrovnik and Rochester campuses had often remained loose. While each campus independently sought a diverse, internationalized student body and faculty, and study abroad was promoted, interaction among students and faculty at the two campuses as part of academic coursework remained infrequent, until recent years. As one example of increased interaction, heightened interest, and greater investment in online learning opportunities, the first year writing course described in this chapter was designed to bring faculty and students at the two campuses together into a globally networked learning environment. First during the spring term of 2012, and again during the subsequent fall term, two sections of the required first-year writing course offered in Dubrovnik and Rochester were electronically linked using a shared learning technology, MyCourses (Desire2Learn). Logistical challenges were minimized because both campuses use English as the language of instruction, share a common curriculum, and utilize a common online learning environment. For ten weeks, students participated in online discussion and collaborative writing projects while enrolled in first-year writing sections offered by RIT professor David Martins and RIT Croatia instructor Rebecca Charry Roje, and supported by RIT instructional designer Michael Starenko.As will be discussed in more detail below, the curriculum for this course was intended to highlight the issues of language difference, which in turn would uncover issues of social justice through students' online discussions and participation.

In this chapter, we explore this international collaboration to highlight productive tensions and instructional opportunities involved in the expanding, transnational character of higher education. First, we explain the aspiration for social justice embedded in our curriculum and offer a brief history of the institutional, curricular, and pedagogical partnership between the two institutions. In that context, we describe ongoing faculty efforts to create high quality learning environments for students, focusing specifically on the online discussion forum and on the asynchronous peer collaboration on writing used in both iterations of the course. We pay particular attention to 1) the ways students used the open discussion forum to share with one another their experience and understanding of culture and language, 2) how the online discussions were integrated into face-to-face class discussions, and 3) the challenges of negotiating meaning across language and cultural difference while students collaborated on writing during peer response activities. Finally, we discuss the opportunities and challenges of facilitating authentic communication between students of different economic, linguistic, cultural, and ethnic backgrounds. Ultimately, through our exploration we have come to realize numerous ways our collaboration has not yet produced the kind of learning we strive for, but we have identified strategies for how we might come closer to achieving our aims.

Language Difference and Social Justice

Language difference is one immediate implication of transnational education, and one that has received critical attention within the field of Writing Studies. Disciplinary discussions have focused on "monolinguistic ideologies" in the teaching of college composition (see Horner and Trimbur 2002; Shuck 2006; Horner and Lu 2008; Horner, Lu, and Matsuda 2010; Horner, Lu, Royster, and Trimbur 2011; Canagarajah 2002). Horner and Trimbur (2002), for example, challenge teachers of writing to reconsider how they respond to language difference, arguing, "an emphasis on the foreign character of students' language has sometimes led not simply to believing in but to attempting to teach a reified version of academic language at odds with its heterogeneity, fluctuating character and negotiability" (p. 620). Because globally networked, online courses link diverse populations of students who perhaps might not otherwise have an opportunity to collaborate together in task-oriented, multi-modal, mediated environments, such courses offer particularly compelling learning opportunities. In particular, students in globally networked classes are continually faced with the need to negotiate meaning in written forms of academic writing, and so gain a personal, nuanced understanding of the heterogeneity and fluctuating nature of language in written academic forms. Min-Zhan Lu (2004) also highlights the relations among language, writing instruction, and social justice. She argues that to be responsible and responsive users of English, writing teachers must:

> combat a commodity approach to the relationship between English and its users, one in which the acquisition of a language, whether a standardized or peripheralized English, is associated with the image of someone first buying or inheriting a ready-made, self-evident, discrete object – a tool (of communication) or a key (to success) – and then learning to use that object like an expert. (p. 25)

This argument is based on evidence that specific forms of English are seen as a commodity that some users acquire and others do not, at least to the degree that their use of English is no longer marked as *other* or *foreign* and therefore not *deficient*. Recognizing that users of English have multiple, sometimes conflicting desires motivating their language use, Lu's strategy for challenging any approach that privileges one form of

English over all others brings attention to "the actual discursive resources of individual writers" (p. 27). Doing so, she argues, encourages writing teachers to design instruction, as we have attempted to do, that draws from students' experience with language use and empowers them to be more effective users of language without privileging one form of English over others.

Bruce Horner, Min Zhan Lu, Jacqueline Jones Royster, and John Trimbur (2010) articulate an approach to language use that does not erase language difference and call for a new disciplinary and pedagogical schema. The authors counter what they term as "traditional" and "accommodationist" approaches to language difference with a "translingual approach" to language difference that "sees difference in language not as a barrier to overcome or as a problem to manage, but as a resource for producing meaning in writing, speaking, reading, and listening" (2010, p. 303). As a practice of social justice itself, a translingual approach to language difference in a writing course not only allows but also elevates for discussion uses of English and ways of using English in written forms. Globally networked learning environments expand meaning making activities for students and provide opportunities for students to develop stronger cross-cultural communication practices.

"Writing Seminar": A Globally Networked Course

When embarking on our partnership, Charry Roje, Martins and Starenko immediately saw the potential for engaging ourselves and students in a more dynamic, cross-cultural and cross-language experience in Writing Seminar, the required first-year writing course that emphasizes academic discourse, research skills, writing and revision. It is traditionally offered to first year students at RIT. In Dubrovnik, where the vast majority of students are foreign speakers of English, the course is offered to second year students who have successfully completed a Basic Writing course in their first year.

Writing Seminar typically focuses on issues of rhetoric, composition and research. We added increased intercultural awareness as a goal of our globally linked version of the course. Included among the student learning outcomes normally listed on course syllabi, our learning outcomes indicated that students would directly address issues of language difference and the varied sources of discursive resources:

- "Students will recognize, analyze and synthesize cross-cultural and cross-linguistic differences, and use those differences as available means for producing meaning in writing, speaking, reading, and listening."
- "Students will identify and analyze a range of influences – social, linguistic, cultural, technological and national – that have shaped them as writers, students, and individuals."

Adding these learning outcomes was based on our belief that by directly engaging with one another in writing, students would have to negotiate meaning with each other in academic discourse. In the process, reified notions of the homogeneity of language and academic discourse could be challenged directly and indirectly.

In planning the course, we agreed that assignments would focus on academic discourse and literacy development and that it was essential to create an environment in which students would feel welcome and safe to explore issues related to intercultural competence and awareness. For example, students were assigned academic journal articles to read and analyze rhetorically, and use those articles as potential models for their own writing later in the course (articles we used include Hawisher et al., 2006, Pandey 2006, Penrose and Giesler 1994). They also wrote a literacy narrative. The "Literacy Narrative" assignment was designed to be a written document developed over a sequence of linked assignments asking students to engage in a variety of writing, talking, reading, and reflecting activities in order to develop their understanding of the social, linguistic, cultural, and national influences that had shaped them as writers, students, and individuals. Ultimately, the purpose of the assignment sequence was for students to reflect upon their own literacy learning as well as that of their classmates.

Because we knew that collaborative, intellectual work on potentially intimate experiences would require close interpersonal relationships, we sought to create opportunities for more informal conversations, specifically in an online discussion forum. We felt that the topics and discussion prompts needed to be broad enough that students could also move the conversation in other directions if they chose. We believed that issues of intercultural communication would become subject to discussion by the fact that we were asking the two groups of students to interact, in writing (Café Bar discussion forum), about writing (peer review), and about academic literacy (essay assignments and course readings).

During the planning stages for the course, Charry Roje was particularly concerned about potential for tension, particularly regarding differences in culture and language. Based on her 10 years of experience teaching Croatian students in an English language environment, she was concerned that that perhaps the Dubrovnik students, who were foreign speakers of English, would be embarrassed or uncomfortable communicating with native speakers, particularly in an online environment where written posts would be public records of grammar and spelling errors. She was concerned that the American students might be seen as a "privileged" group, both due to their status as native speakers and cultural assumptions on both sides about American geopolitical influence. Based on her own experience of living in Croatia for 10 years, becoming fluent in Croatian language, Charry Roje also wondered if differences in culture, or stereotypes of "others" could lead to an "incident" in which some students might unintentionally offend one another. In other words, both language difference and cultural differences seemed to provide potential opportunities for conflict, as well as for learning.

During instruction, however, differing levels of English competency did not seem to become a source of tension in the group and did not surface in public discussions. Additionally, a number of the Rochester students were also foreign speakers of English, and some of the native speakers did not appear to be overly concerned with grammar or spellchecking in their posts. Because the course did not focus on traditional issues of language competence (such as grammar, spelling, syntax), students seemed not only to accept but to disregard both their own and others' "errors" in standard English usage.

In what follows, we explore the students' use of the online discussion forum, the issues of social justice and language use that were raised when the online discussion was integrated into face-to-face class discussion, and the peer review activities that highlight the challenges of negotiating meaning across difference.

Cultural Sharing in Online "Café Bar" Discussions

We named the online discussion forum the "Café Bar: A Virtual Space for the Culturally Curious." During the first iteration of the course, only one of the six discussion topics posted to the Café Bar directly prompted reflection on issues of cross cultural communication. In the second week of classes, after introducing themselves in a post entitled "Welcome to My World," students were asked to share their "Cross Cultural Resume." This second prompt asked students to detail their previous contact and experience with "various cultures and languages other than your own, perhaps related to your family background, travel, friendships or reading." None of the other discussion topics prompted reflection on issues of culture directly, but rather offered students the opportunity to get to know each other better by sharing photos, recommending music videos or books, or describing hobbies. We expected that issues of cultural and language difference would surface as a result of these conversations, but we were more concerned that these low-stakes interactions establish a tone that would provide students familiarity with one another so that they would feel comfortable working with one another on the more academic writing tasks.

To encourage open communication, discussion posts were assessed on a credit/no credit/resubmit scheme. Students received full credit no matter what they wrote if they submitted thoughtful posts according to announced deadlines and followed instructions. We believed that this assessment strategy would alleviate student anxiety about grades and foster a more open and honest conversation. In practice, the students showed a high level of participation as shown in the number of posts produced by students in both offerings of the course. Interesting patterns of response will be discussed below.

When introducing themselves to the group, several students used the Café Bar, to raise issues of social justice. While some students in both sections chose noncontroversial topics such as summer vacations, others demonstrated an interest in the wider world around them. One female student shared an image of the men's rugby team that she played on, describing her experience being a woman playing a stereotypically male sport at a traditionally male-dominated institution. This student wrote about the percentage of female students at RIT, and noted the small number of women enrolled in her section of Writing Seminar. Another student in Dubrovnik responded with stories of her own experiences as a woman in male-dominated martial arts practice. A third student, from Rochester, described his experience as a member of an ethnic minority overcoming disadvantages in high school and now college. In addition to these issues regarding diversity in US higher education, one student from Croatia wrote about a controversial gay pride parade that had been held in Split, Croatia. Other students wrote about female participation in college athletics, environmentalism, homelessness, consumerism, and humanitarian responses to the earthquake in Japan.

In the "Cross Cultural Resume," students in both groups discussed their family roots and significant life experiences with different cultures and languages, including Ghana, Vietnam, Korea, Spain, Germany, inner-city Rochester, and New York City. The titles of the Café Bar posts written by students at both campuses were written in a variety of languages – French, German, Croatian, and English, and one suggested a hybrid language: "SpanCroGlish" – and also focused on issues of language and culture: "Do you speak English?," "Culture meets Culture," "New Cultures and languages," "Rediscovering a lost culture," and "Ragusa Language."

Even with all of the attention on broad-based linguistic and cultural experiences, five of the 16 students in the Rochester class identified themselves as not having much cross-cultural experience and/or identified themselves as somewhat impoverished in this regard. These were students whose experience was limited to brief vacations or foreign language study in high school or who felt estranged from immigrant family roots over generations. When writing about his limited experience one student wrote: "I cannot honestly say that I'm a particularly culturally knowledgeable person […] which is actually kind of sad." Other students wrote about their cross-cultural as "not much to say" or the "bare minimum," or that they were "sheltered culturally," and their lack of experience was "unfortunate." There seemed to be a general consensus, however, that cross-cultural awareness and experience was to be desired, and that students could only benefit from such experiences.

Based on our experiences during the spring term, we decided that the issues being raised in the Café Bar were too generative to leave as informal writing meant primarily to improve the relationship among the students in the class. For that reason, we altered the Café Bar prompts to address specifically the issues raised in course readings. Because we didn't alter our own approach to responding to the posts, we again observe numerous missed opportunities to bring the discussions occurring in the Café Bar into broader class discussion.

The fall offering of the course started out similarly to the spring, with *icebreaker activities* meant to acquaint the two groups that included creating and exchanging introductory videos of each group, which students seemed to enjoy. In response to the videos presented by the students in Dubrovnik, the Rochester group created a second round of individual introductory videos which they posted on a YouTube channel they created for the exchange (see YouTube Channel "COILWritingSem"). Participation in the first café bar discussion was high, with students posting photos and commenting extensively (see Table 1). There were 22 original postings, and each student was required to response to two postings. Because the numbers of students in each class were so different – 16 in Dubrovnik and only 6 in Rochester – the participation of the instructors were also high. Charry Roje and Martins, for example, wrote a combined 29 responses to student posts.

Coming at the close of the tourist season, the fall offering of the course brought with it a number of "Cross Cultural Resume" posts focused on the significant role of tourism in Dubrovnik. Because of the 6 hour time difference between the class meetings, every one of the students in Dubrovnik posted their initial thoughts on the prompt before the students in Rochester. This accident of time difference and the prompt itself led to numerous posts on travel and tourism experiences from the Rochester group as well, with students often focusing on obvious cultural differences they experienced while traveling to other countries (e.g., traditional hair styles in Belize, and meat-based diet in Nigeria). Even so, students didn't focus solely on tourism, and a variety of provocative statements about language and cultural difference were made.

Highlighting the benefits of difference, one student from Croatia, wrote about having learned about other languages and cultures as a child: "Now when I am a student, I realize that knowing other languages or understand new cultures you can spread your own horizons about life." Another Croatian student also wrote about the limitation of not needing to understanding different languages or cultures: "I think that if we don't have need to understand a different language or culture that we are limited people, and that our life is a very small and poor life." One student from the U.S. responded by pointing out what she saw as the limited opportunities in the U.S. of learning and using other languages:

> In America, we are not given a lot of great opportunities to learn new languages efficiently. We are taught that in every county where English is not the native language, the people have the opportunity to learn it well and they usually do. So we think "Oh that means we can communicate because they learned it, we don't have to work hard for it." It's a horrible thing.

The Croatian author of the original post responded, "It would be great if we all spoke the same language, but I agree with you that learning a new language is hard. However, it is good feeling when you know some other language because you can understand other culture." Unfortunately, this exchange, which could have been the basis for a rich discussion of systemic, cultural, and geographic influences on language learning and of opportunities for using a second language, remained limited to two individual students. Although students were all drawing on their experiences and highlighting important factors that influence the development of actual discursive resources, as instructors we inadvertently positioned ourselves – because of the social intent of the discussion forum, and because we required each student to write three posts – to miss the learning opportunity of exploring these specific issues within a whole-class discussion. As the instructor of the U.S.-based section, for example, Martins missed an opportunity to have a discussion with students about the limitations of monolingualism, or the implications of their self-referential move to identify as culturally limited.

Another example of an exchange that stayed "local" to a discussion between two students highlights one of the key moments when this collaboration between Rochester and Dubrovnik might significantly raise issues of social justice. In what follows we present longer sections of discussion posts written by the students in order to highlight the complex issues they raise as well as the language they use. During this short exchange, a Croatian student describes his first cultural experience when he and his family were forced to Italy during the Croatian War of Independence:

> My first cross cultural experience, as far back as I can remember, happened during the Croatian War of Independence. My mother, sister, and I had to flee the town, so during our time away we moved a lot. One of the towns I lived in for a few months was a town in northern Italy called Udine. [...] Anyway my first cross cultural contact happened there. I encountered a different language for the first time. For the first time in my life I was outside of Dubrovnik and it all seemed so strange, but, yet, so exciting because as a kid it all looks fun.

As a consequence of the rebuilding of Croatia after the war, the student ends the post by describing his experience living and working in a town that is a tourist destination:

> I am glad that I live in a town that is a tourist destination. You can always get in a conversation with people from other parts of the world and learn the culture info first hand. I actually enjoy it, and that's why sometimes I like working in tourism. But on the other hand, I get tired of dealing with strangers and just want to relax.

Astonishingly, rather than asking questions about fleeing a war, the one response came from a Rochester student who focused on the benefits of interacting with people from other countries who speak different languages:

> Although we don't get a lot of interactions with tourists in Rochester, I believe the diversity at RIT relates with your interactions with tourists. Everywhere you look, you can see a person of a different ethnicity or culture here. I love this because, just like how you said, I have the opportunity to always meet someone new and different and learn about a culture that I barely know anything about. Everyone can go online and Google information about a culture, but talking with someone of that culture is so much more interesting and allows room for so much detail that the internet couldn't provide. This diversity is one of my favorite aspects of RIT.

Through her response, this student draws attention to how discursive resources are developed – through direct interaction with others and internet exploration – and she suggests her own utilization of such resources by describing her love of meeting new and different people and learning about them. Other issues, such as diversity on a college campus, are also rich topics for discussion, especially if we were to address the legacy of English Only language instruction, systemic racial and gender discrimination in U.S. higher education, or the disproportionate allocation of resources for primary and secondary education along racial and class lines. However, because of our objective to allow the Café Bar forum to be a safe space for students to exchange experiences, ideas, and views, some of these incredibly rich opportunities for further exploration were missed by both students and instructors.

Generally speaking, as the examples above demonstrate, we saw that some students in both sections came to the course with an already established interest in potentially controversial topics related to social justice, and felt free to express opinions which could be considered provocative. Their posts could have invited the kind of larger discussion that would enable students to investigate and even question some of their own cultural values. However, these posts were rarely responded to in the online environment. Rather than producing useful discussion, the online forum tended towards what we have come to think of as "Pleasantville." The silence in response to these posts suggests that students were reluctant to disagree, or rebut, instead focusing on creating an atmosphere of polite civility. We understand this to be the result of students' concern about presentation of self in a public forum to their peers and their teachers; that is, especially in these discussion posts early in the class, students are presenting their best selves, engaging in polite conversation and demonstrating to their teachers that they can effectively complete the assigned tasks which require response to issues raised by their peers. While there is clearly some risk avoidance within the written posts, students in both sections openly expressed interest in some of the more substantial issues addressed above. The task for us as the instructors is to facilitate more substantive reflection on such topics through the structure of assignments, our own responses to discussion posts, and class discussion.

Integrating Online and F2F Discussion: Regional Differences in Croatia

Compared to these somewhat limited online exchanges, there were moments when the discussions in the Café Bar forum were integrated into the face-to-face classroom discussion. Interestingly, those moments tended to be initiated when a topic was being discussed online among students from one class. In both offerings of the course, for example, students in Croatia responded to the "Cross-Cultural Resume" prompt by writing about the regional differences they observed while living and traveling in different parts of Croatia. During the fall offering of the course, this online discussion was so compelling for the Dubrovnik students that it prompted multiple posts discussing stereotypes of American, Italian, Russian, Asian, and Croatian people. With over 35 posts on related issues, the online discussion was so energetic that Charry Roje devoted one full class session how "othering" happens when people rely on stereotypes to understand cultural difference.

The discussion started when one student posted a lengthy but nuanced comment noting differences between cultures in Northern and Southern Croatia. Identifying himself with the Southern group, he voiced negative stereotypes about Northerners. However, he was also careful to note factors that might have biased his point of view, and invited others to join the discussion and clarify. Again, in order to highlight students own language, we present a significant portion of the original post:

> From what I've seen, the people in the south of Croatia are [different from] people who live in the north. They are more friendly, more open, but not too open, and more fun people in general. However the people in the north are, as I said, different. [...] People seemed a lot colder, more reserved, definitely less friendly. I remember what my parents would always tell me: watch your pockets or someone might steal something. I think that is the reason why people there are different. They're always on guard, they are careful, not trusting. [...] But again, I was never there for more than 10 days in a row, so maybe I didn't see enough. I would like to see more opinions about people there, as I might not know enough.

Other students quickly responded, agreeing that they too had noted north-south differences. Students seemed interested in exploring these differences, and posing possible explanations, and most seemed careful not to assume too much about the cultural differences that had been observed or experience.

Soon, the online conversation expanded to the use of English in Europe, with students posting comments that turned the discussion away from their "own" culture to a neighboring one. This issue of "othering" a group of people based on common stereotypes became prevalent in other exchanges written by the Dubrovnik students. In an unrelated post, for example, another student posted a comment about how his trip to the U.S. confirmed his view that "America in live view is no different from the one presented on a TV." An American student studying in Dubrovnik responded to this comment with concerns about making generalizations after only a week of experience. Charry Roje, the Dubrovnik instructor, echoed these concerns and shared a related experience as an American coming to Croatia with the intention to validate the student's feelings by sharing a similar experience, and then move the conversation toward how initial negative reactions can be more deeply examined.

Interestingly, none of the Rochester students responded to the original comment, which obviously could have been taken as insulting or offensive. Based on the Café Bar exchanges and in-class discussions, the Dubrovnik students seemed eager to discuss cultural difference with many of their posts oriented toward issues of negative stereotypes. While some students cautioned about the dangers of stereotyping, others openly disagreed: "I would disagree with you. The stereotypes are correct and useful. I am saying that because I have never met a quiet and polite Italian, they are all rude and loud. English speaking Russians or Asians, they are always the same." These posts precipitated a lively in-class discussion about whether stereotypes are useful.

During the face-to-face session of the Dubrovnik class, Charry Roje focused explicitly on the contrast between the "pleasant" posts, which, as a group, seemed to find only positive things in cross cultural interactions, and a second group of posts which seemed to express mostly negative sentiments about "others." Students were asked whether they thought this was a contradiction, and if so, how they would explain it. Although the conversation was not planned in our original schedule of activities, Charry Roje used the well-known analogy of an "iceberg" to introduce students to the distinction between *observable behavior* – the tip of the iceberg – and unseen *cultural values* – under the surface of the sea. At Dubrovnik, the in-class discussion focused on potential elements which *bias* our points of view and taint our interactions with others. In Dubrovnik, dependence on tourism means that even those who work outside the tourist industry have a great deal of contact with people traveling to Croatia. The class argued that interactions with others are too often tainted by economic exchange, with locals aiming to earn money and visitors suspicious of being taken advantage of. After some sharing of stories, students agreed that "the money factor" would cause tourists to "behave differently" and locals to "see them in a different light."

The student from southern Croatia who authored the comment about northerners was prompted to consider whether any contextual factors may have influenced his conclusions. During the class discussion it was suggested by others that his impression that southerners were friendlier could have come from generalizing his own experience as a local in his small hometown; coming from a small town, he could easily feel that people were "less friendly" in a city of 1 million where he was a stranger. The student who was disappointed in his American experience was encouraged by his instructor to say more about what exactly happened that led him to his conclusions, and was encouraged to choose some aspect of behavior that he observed in America (and which bothered him) as a topic for his research paper.

The majority of the class period was devoted to discussing these "controversial" posts debating the value of stereotypes, and exploring the reasons why forming generalizations about others is so tempting and seems at times useful. Students were very engaged despite the apparent tension in the online discussion. In an informal, anonymous course evaluation immediately after the in-class discussion, students were asked to evaluate, among other aspects of the course, the class discussions. On comment reflects a common sentiment: "Class discussion is actually the part where I learn the most." Perhaps as a response to the discussion earlier in the day, another student wrote, "I have learned to tolerate other opinions." Such comments illustrate a positive moment in cultural dialogue, in which students felt free to voice honest, negative impressions of others. As a class, students were able to reconsider their views in discussion with peers and the instructor. The fact that this generative discussion happened only within the contexts of the Dubrovnik class highlights the benefit of offering blended instruction, which allows for both online and face-to-face conversation.

Rochester students were also asked, "What have you learned (e.g., about online discussions, about cross-cultural communication, about other places in the world, about yourself, etc.) doing the Café Bar discussion posts?" Nearly all of the students highlighted their keen interest in learning "about other cultures and people's backgrounds" and "how things are done in different places," but they shied away from the more difficult tasks of working through cultural and language difference. One student described the difficulty of online discussions: "In terms of online discussion, some translation gets lost: when having a conversation in person with someone, it's easier to ask questions while they're talking about something for clarification. Online that luxury is lost and some things aren't as clear as others in what the writer is trying to say." Another student focused on specific genres of writing that she preferred: "I do enjoy writing about what I know, and anything else I'm plain bad at. I'm much more confident when it comes to narratives on me or something I know well. Other ways of writing I'm not good at." Both comments, we believe, reflect how difficult collaborating on negotiating meaning across cultures and languages can be, especially online. Still, the issues students are raising in their online discussion forum highlight the importance that such negotiation of meaning can have for improving cross-cultural and cross-language communication competence. In our experience it has been the integration of the online discussion in combination with face-to-face discussion that has led to serious, even mind changing, dialogue.

"My Way or the Highway": Negotiating Meaning in Collaborative Writing Tasks

Blended offerings also make possible situations where stereotypes one group has about the other can crystalize. While the previous section illustrates opportunities for negotiating meaning *within* a cultural group, perhaps the most challenging moment of all occurred *between* the two sections during a collaborative writing task in the fall offering. This last example from our class illustrates an essential difference between cultural issues addressed directly in online discussion and those that lie *under the surface*, arising indirectly in the process of working collaboratively on a written project. Issues of plagiarism, authority, and representation of self all came to the surface dramatically when students were engaged in peer review. The exchange highlights the strengths of the discursive resources students and instructors exhibited, and draws attention to the broader range of resources that this type of course endeavors to cultivate.

As part of the assignment sequence meant to help students identify and analyze the range of influences that have shaped their own development as writers, students, and individuals, students interview each other and wrote "literacy profiles" of two classmates. At the time of the exchanged described below, students had already worked through the sequential, collaborative process within the same section, where interviews were conducted face-to-face, in students' native language, among students who share similar cultural backgrounds. For the second profile, students were asked to collaborate across sections, interviewing each other online, writing a literacy profile about the peer from the other class, and then providing peer response on the document in which they were profiled.

During the fall, because of the disparity in class sizes, each Rochester student interviewed and profiled one Dubrovnik student, and small groups of Dubrovnik students interviewed one Rochester student. In-class discussion identified the challenges of such group interviews, with particular attention to the difficulties of online, asynchronous interview conducted over email. Students were encouraged to "get to know each other" before the conducting the actual interview, using live chat or Skype to discuss hobbies or other low risk topics. They were also encouraged to follow up on written answers to questions before writing their final drafts. However, in practice, students completed this interview process on their own, outside of class, without synchronous communication, and without direct instructor supervision or oversight.

The first round of peer review called for "corrective feedback," allowing the subject of the written profile to offer any feedback that might correct any errors or offer further details about the events described in the profile. In their responses to the profiles written by Rochester students, all six of the Dubrovnik students accepted the profiles that were written about them with very little comment. One student added a minor factual clarification, but all others simply wrote brief statements of approval in reply messages on the comment board. For example: "Profile is just fine, it is very accurate" and "I am perfectly satisfied with this."

In contrast, some Rochester students felt they had been misrepresented or misunderstood personally in the profiles written about them. The most obvious of these was a profile of one Rochester student, who at some point during the interview process, had used the phrase, "my way or the highway" when discussing her early experiences with group work in school. The Dubrovnik students who interviewed her chose this phrase as the title for their profile of her: "Susan B.– My Way or the Highway."

When she saw the title, "Susan" responded with the in-text comment: "Why this as the title? When first reading it, it makes me sound like a completely different person than I am. My way or the highway means a person who gets mad when they don't get their own way and are very stuck up about it." Although she seems to feel clearly misrepresented, the student did consider that perhaps the problem was in understanding the colloquial idiom. However, as later in-text comments show, she was also upset. One comment highlights a sentence with the comment, "Not true," followed by "not true at all" about another passage. Later in the draft, the student highlights passages in green and other passages in red. At the end of the paper, she explains that the red passages identify those parts that were not true, or did not reflect what she had said in the interview. The passages highlighted in green identified the issue of "copy paste" and potential plagiarism: "Most of this is written by taking my answers straight from the interview questions and changing the pronouns, I feel like I wrote this piece, not your group."

Because of the time difference, Charry Roje was able to alert Martins to the situation before he met with the Rochester students. Martins was then able to spend half that morning's class period discussing this particular peer review and ways of making peer review more productive. As a result of the early intervention, the student was able to revise her comments and the other five students in the class who had not yet responded to the profile written about them were primed to respond in more productive ways. In their responses, the Rochester students offered a range of comments on the texts that had been written about them, annotating the original texts with comment bubbles (in Microsoft Word), color-coded highlights, and other forms of in-text commentary. While some comments were basic "fact checking," others were more oriented towards peer review, and offered organizational strategies for revising the paper. Several RIT students even labeled their "corrective feedback" documents as "editing" or "peer review" in the file name. One student attached her file with the comment: "Here is my *revision* of your profile" (emphasis added). Another student offered comments on how to rewrite the paper, using phrases like "You should expand on…" and "You should explain…" and a final comment with explicit directions for rewriting: "You really need to go back and expand on everything so that it connects back to my literacy development. Don't just say what has happened in my life but also explain the importance of it happening." It is impossible to know how they would have responded without the class discussion on the first student's response. However, because the initial response occurred, the class was able to explore an authentic experience of negotiating meaning across cultural and language difference, use that exploration to develop their own discursive resources for negotiating meaning in writing.

In Dubrovnik, during the class meeting immediately following this exchange of online feedback, students were noticeably upset. As class began, one student volunteered casually, "It's funny that in a class that is supposed to make us more friends with the Rochester students, it ended up making us not like them." Sensing the tension, Charry Roje asked students to write down their general reactions to the feedback they had received from their RIT classmates. This feedback was intended for the instructor only, and students were told it would not be shared with the Rochester group. While students whose profiles had received only minor fact checking corrections seemed to feel that the process had been "good" or "easy enough," students who received extensive feedback beyond fact checking from their peers were dissatisfied and felt that their own creative process and even authority to write the paper had been undermined, and that the Rochester students were just "too sensitive."

One Dubrovnik student even went so far as to write "young students at [the Rochester campus] have a lot of problems with themselves. We are much more relaxed about what people say about us because we know who we are, and they are so worried about things others think about them. They can't accept who they really are." It is important to note that this particular, very strong reaction was not typical of the group. Most other Dubrovnik students noted that the very limited, asynchronous communication and lack of context may have contributed to the problem. However, this "my way or the highway" incident did seem to cool relations between the two groups significantly. After this point, we did not fully recover the positive atmosphere with which our interaction began.

In the next class discussion the students in Dubrovnik were encouraged to think about whether they had actually misunderstood or misrepresented their Rochester classmates, or whether issues of language difference had been at work. As a group, they seemed to resist self-criticism and continued to blame the Rochester students for being too sensitive. In fact, one student from the "My Way or the Highway" group was even more convinced after this exchange that the title was appropriate in describing the Rochester student. In this way, experiences of cultural and language difference can also lead to stronger group identification.

While the Dubrovnik students were reluctant to criticize themselves, they were more open to criticizing the conditions of the assignment which might have led to conflict. When asked to provide details about how the interviews had been conducted, many of the students confessed that they had not conducted any synchronous discussion, either before or after the formal interview itself, and relied strictly on asynchronous email exchanges of written answers to questions. When asked by the instructors, students recognized that this process was a possible reason for the "misunderstandings" that later surfaced in the profiles. Although these in-class discussion seemed to diffuse some of the initial negative feelings, students did not ask for any cross-sectional follow-up activities and enthusiasm for the exchanges between the two classes decreased somewhat.

Charry Roje and Martins discussed ways to handle the situation, but did not intervene. Instead, the instructors presented a Café Bar prompt on the topic of authority in writing. Students were asked to respond to a course reading by sharing a story about a writing experience that draws attention to "the factors that you believe led to your sense of authority, and how it effected what you did as you completed the writing task. Tell your story, with detail. (It doesn't have to be about writing an essay. Look to your profiles, too, to find stories to elaborate on…)." One student from Dubrovnik who had been vocal in class about how the peer review had interfered with her creative process wrote a post titled "Respect my authority":

> I get pretty annoyed when someone corrects my writing just because it's not written in their way. I will always accept constructive criticism [sic] if I make grammatical or factual error, but what I write is completely up to me. The most recent event that i connect with this is our profile essay feedback, I really don't mind critique based on facts but I mind people trying to change my view of them.

This comment – one that raises important issues of authority in writing, the scope and nature of feedback, and ideas about the role of dialogue – was not responded to by any of the Rochester students. One Dubrovnik classmate did respond, but simply voiced general support. After a rather dramatic and somewhat difficult asynchronous confrontation, students seemed to return to "Pleasantville."

Opportunities and Challenges

Issues of social justice seem naturally to arise in the writing students produce for a course that asks them to read and write about their development as students, writers, and individuals. Yet, in a course that linked students from different parts of the world and asked them to collaborate online, we found that moments for genuine self-reflection, and re-examination of cultural beliefs and notions of others, were often missed. In addition to the technical and timing issues of globally networked learning, differences in language, culture, and schooling present two distinct, and in fact, opposing, dangers: on one hand, students may avoid controversy entirely, while on the other, they may misunderstand and offend each other and end up reinforcing existing stereotypes.

Our experience teaching two globally networked courses demonstrates that an asynchronous online discussion board, in isolation, clearly tends toward a sort of sterile "Pleasantville" in which many students shy away from genuine opportunities to re-examine or question established cultural attitudes. The discussion board format asked students simply to "chat" virtually, for more social purposes. Our students tended to see these low-stakes online forums as places to present their public selves, and were in general reluctant to question or confront others, or to present ideas or values that could perhaps risk offense.

Although students seem somewhat less reluctant to directly disagree with peers in their own local face to face group, students tend to avoid risking confrontation with students in the "other" group, at least in the context of low-stakes, online discussion. Although potentially serious, sometimes controversial, issues of cultural and language difference did surface in the online Café Bar, most students seemed reluctant to pursue these opportunities, and preferred to keep the discussion in a "safe" zone. Therefore, as instructors, our task going forward will be to design activities that make students feel safe taking the risk to explore and engage more deeply in what their classmates have written. A key to these activities may involve emphasizing more synchronous communication, perhaps involving audio and video in addition to writing, to help students establish deeper relationships before writing.

The challenge to making online discussions more productive is to facilitate more substantive responses to cultural and language difference through the deliberate structuring of writing activities, focusing our own responses to students' writing, and utilizing the *blended* nature of our courses (i.e., both online and face-to-face meetings). With so many original posts produced in each class, the number of "discussions" that raised the issues we hoped to explore was overwhelming. We missed numerous opportunities to engage students in writing about those issues. By managing the discussions, referring to them more directly in face to face classes, and revisiting discussion threads more deliberately, we can provide students opportunities to learn from one another, and gain a deeper understanding of the issues the collaboration brings into the foreground. For example, after students introduce themselves in their "Welcome to my world" post, we could form the students into four different groups for the "Cross Cultural Resume." While the first group might post an initial response to the prompt, a second group of students could provide what David Baker calls "substantive comment" that offers "a well-constructed, thoughtful, independent comment" to the original post (404). A third group making a "responsive comment" that offers a "thoughtful, independent but shorter comment to the substantive comment of another discussant" (404). The final group of students could then be tasked with relating the ensuing discussion to course readings and in-class discussions. Regardless of what roles the different student groups may play, asking students to do more than simply respond to two of their peer's posts would help us all think more deeply and learn more about negotiating meaning from positions of cultural and linguistic difference.

In contrast to our "Pleasantville" experience with online discussion boards, in which students simply "conversed" with each other and avoided confrontation (and therefore opportunities to expand their cultural awareness), assignments which required students to actually collaborate in producing a piece of writing together, seemed to bring underlying issues of language and cultural difference to the fore. We witnessed potential for misunderstanding, conflict and even reinforcement of cultural stereotypes of others, when students collaborate on written assignments in which they may not share the same underlying assumptions about authorship, and authority. More importantly, our experience points out the significance of local cultural assumptions about education and students' previous educational experiences, for how students respond to assigned writing tasks and to one another. As our "my way or the highway" incident clearly demonstrated, these interactions have the potential to strengthen, rather than correct, the kinds of stereotypes of other cultures that we hoped our course would address.

Finally, while it is comparatively easier and safer to write directly about cultural issues in the informal context of an online discussion, it is quite another matter to negotiate meaning across cultural and language when they surface unexpectedly during other course tasks like peer review and collaborative writing. Given the aims of our course, we see these moments of tension or conflict when working collaboratively on a project, not necessarily as problems to be avoided, but as rich ground for further exploration of cultural and language differences which may be hidden under superficial similarities. For this reason, working with students to reflect meaningfully on these issues within online discussions will certainly help students negotiate differences when working together to produce writing. By attending more deliberately to the structure of online discussions, expanding in-class discussions to include both sections, revising our assignment structures and instructions, and publically reflecting on the writing process throughout the course, we are confident that students can come away from globally linked learning experiences with deeper understandings of other cultures as well as their own.

References

Baker, D. L. (2011) "Designing and Orchestrating Online Discussions." *MERLOT Journal of Online Learning and Teaching* 7(3): 401–411.

Canagarajah, S. (2002) *The Geopolitics of Academic Writing*. Pittsburgh: University of Pittsburgh Press.

Hawisher, G. E., Cynthia L. S., Yi-Huey G., and Lu L. (2006) "Globalization and Agency: Designing and Redesigning the Literacies of Cyberspace." *College English* 68(6): 619–636.

Horner, B., and Trimbur, J. (2002) "English Only and College Composition" *College Composition and Communication*, 53 (4): 594–630.

Horner, B., and Lu, M. (2008) "Resisting Monolingualism in 'English': Reading and Writing the Politics of Language." In: Ellis, V., Fox, C. and Street, B. (eds.), *Rethinking English in Schools: Towards a New and Constructive Stage*, pp. 140–157, New York: Continuum International Pub. Group.

Horner, B., Lu, M-Z. and Matsuda, P. K. (eds.) (2010) *Cross Language Relations in Composition*. Carbondale: Southern Illinois University Press.

Horner, B., Lu, M-Z., Royster, J. J., and Trimbur, J. (2011) Opinion: Language Difference in Writing—Toward a Translingual Approach. *College English* 73(3): 303–321.

Lu, M-Z. (2004) "An Essay on the Work of Composition: Composing English against the Order of Fast Capitalism." *College Composition and Communication* 56 (1): 16–50.

Pandey, I. P. (2006) "Literate Lives Across the Digital Divide." *Computers and Composition* 23: 246–257.

Penrose, A. M., and Geisler, C. (1994) "Reading and Writing without Authority." *College Composition and Communication* 45(4): 505–520.

Shuck, G. (2006) "Combating Monolingualism." *WPA* 30 (1–2): 59–82.

Part III: Intercultural Encounters as Transformative Experiences

Wendy Anderson and John Corbett

'What Do We Chat about When We Chat about Culture?' The Discourse of Online Intercultural Exchanges

The Educational Context

This chapter draws upon seven years of experience in teaching a course called 'Culture and English Language Teaching' to final-year undergraduates at the University of Glasgow.[1] The course addresses issues in intercultural language learning (e.g. Corbett 2003; Risager 2007; Byram 1997, 2008) through a combination of lectures, student presentations, class discussions and experiential online exchanges through 'Intercultural Connections', a specially tailored virtual learning environment. 'Intercultural Connections' allows spaces for students from Europe, South America and Asia to explore aspects of their culture and identity for themselves, and, for the Glasgow-based class (which typically consists of a majority of UK undergraduates as well as European and Asian exchange students), there is also explicit input on the goals of intercultural learning, and ethical, political and practical issues in language education. Students may post online the outcomes of intercultural activities, such as home ethnographies, reactions to literature from other cultures, and sample language learning activities. One discussion forum is set up as a 'café' where students initiate and develop their own discussions.

1 The authors owe a debt of gratitude to all the participants in the 'Intercultural Connections' site over the years, who, by registering for the course, confirmed their agreement that their postings might be used anonymously in research and further teaching. We are also grateful to our colleagues, especially Professor Alison Phipps, Andrea Assenti del Rio, Dr Peih-ying Lu and Howard Miller, for their support in running the course each year. The course development also benefited from the award of Learning and Teaching Development grants from the University of Glasgow.

Unlike most other examples of the burgeoning field of online language courses (e.g. Meskill and Anthony 2010; O'Dowd 2012; and the examples gathered in O'Dowd ed. 2007) 'Intercultural Connections' was devised as a means of enabling both native speaker and non-native speaker students who shared a general interest in teaching language and culture to develop and reflect on aspects of intercultural communicative competence through experiential interactions that would complement more traditional 'lecture and seminar' delivery of content. Following Byram (1997, 2008), intercultural communicative competence is conceived of as a set of resources – competences and skills – that the learner is encouraged to develop in order to facilitate:

- Knowledge of the self and the other
- Knowledge of how to relate and interpret meaning
- Knowledge of how to discover cultural information
- Knowledge of how to relativise oneself and value the attitudes and beliefs of others
- Critical cultural awareness.

Students were encouraged to read about and discuss intercultural communicative competence as part of their year-long course; the online environment was envisaged as a forum for actually doing it. Aspects of the 'Intercultural Connections' platform have been discussed elsewhere by those who established and taught on it. Corbett and Phipps (2006) report on the early findings of the course, raising questions about whether the course lecturers could be said to be teaching the *savoirs* that Byram articulates, or whether the online environment simply allowed a space for interactions to happen that, however erratically, unpredictably and partially, exemplified them. Anderson and Corbett (2012:113) revisit this issue from the perspective of later iterations of the course, concluding that:

> While it may not actually yet be possible to establish firmly one way or the other whether attitudes of openness can be taught, it would seem that the skills of reflection developed through intercultural exchange, in the context of a CMC-based course, with both a theoretical and a practical focus on issues related to the teaching of culture and English language, can certainly heighten participants' awareness of their own attitudes and help them to value the attitudes and beliefs of others.

In the present chapter, we develop our discussion of our experience of this course by considering in greater detail than before a sample of self-initiated interactions by course participants (2004–11). These are complemented by the data from a set of semi-structured interviews with 15 Glasgow-based members of a recent cohort, representing both native and non-native English speakers, which enabled us to probe the participants' reflections on their own experience of online intercultural exchange throughout the course. The interviews were carried out at the end of the academic session 2010–11, and focused on a number of themes relating to the use of the 'Intercultural Connections' moodle site.

In line with the aims of the present volume, we ask what the interactions have to tell us about the nature of intercultural dialogue in an interconnected era. The online interactions discussed below are related to Blommaert's (2010) model of globalised language hierarchies, and to aspects of situated, discourse and transportable identities, as formulated by Zimmerman (1998), Richards (2006) and Ushioda (2011). We also revisit and borrow from the work of Belz (2002), whose discussion of telecollaborative dialogue as a potential realisation of intercultural communicative competence is grounded in the linguistic features of appraisal (see Martin 2000). We follow Belz in recognizing that the appraisal system offers a potential means of tracking affective expressions in a small corpus of computer-mediated interactions, and so offers applied linguists an index of attitudes presented by participants in online intercultural exchanges, particularly as they perform and negotiate their discursive and transportable identities in interactions that blur the distinction between local and global.

Global Discourse and Participant Identities

Blommaert (2010) calls for a new 'sociolinguistics of globalisation' that can account for the collapse of time and space that is characteristic of online communication. To summarise his position, Blommaert first of all sees language in use as governed by situated norms, or 'orders of indexicality' that apply differently to different contexts. The norms governing informal conversations are not those that apply to written job applications, for example. Blommaert (2010: 34–5) argues that the orders of indexicality are sedimented in layers that are governed by images of time and space: an informal conversation would be momentary, ephemeral, local and situated (i.e. on a lower scale), while a written job application functions translocally, according to norms that are more widespread and 'timeless' (i.e. on a higher scale). Blommaert suggests that in producing discourse, speakers may jump from one 'scale' to the other, for example by producing in an informal conversation an expression associated with the norms of written language, or vice versa. While this phenomenon has been variously described as 'intertextuality' or 'register shifting', Blommaert prefers the description of 'scale jumping' to produce language that indexes social order. To invent an example, a group of students in a British pub might tell a visiting exchange student that 'cheers' is a more appropriate expression of gratitude in this local setting than 'thank you for your kind consideration'. They would thus prevent an accidental jump in scale, and socialise the visitor in respect of the local order of indexicality.

All online interactions represent a complex blurring of the local and the global. Online 'chat' has many of the indexical features of ephemeral, informal conversation; however, it also takes on a translocal, widespread function too. As our data represent interaction between Glasgow-based participants and participants elsewhere, there is a tension between the locally-indexed language and the global nature of the interactions. To put it simply, just because participants share a digital platform and a lingua franca does not mean that they find it easy to coalesce into a discourse community. Online chat is saturated with local indices – slang terms, situated cultural references – and translocal participants need to find ways of negotiating these potential obstacles to global communication.

The tension between local and global orders of indexicality also has an impact on the identities of the participants as they are co-constructed through the developing discourse. Ushioda (2011), following Zimmerman (1998) and Richards (2006), identifies three aspects of identity relevant to the language classroom: *situated* identities that would be conferred by the context to the communicative event (e.g. student in an online discussion group); *discourse* identities that relate to the role that the participants play in the discourse (e.g. initiator of a thread, responder to a question, lurker); and *transportable* identities that are invoked by the interaction, e.g. football fan, pet-owner, lover of fish and chips, and so on. In much of the online chat, the Glasgow-based students act as initiators of the discussions, bringing aspects of their transportable identities to the chats as the basis for interaction. The degree to which overseas participants also initiate, respond and negotiate their transportable identities speaks to the degree of their intercultural communicative competence. However, the degree to which Glasgow-based participants negotiate the local and translocal scales of their interaction may well be a factor in facilitating or preventing broader participation in the online discussions. In the following sections, we consider these issues in greater detail.

The platform

In 2005, the University of Glasgow adopted 'moodle' as its preferred virtual learning environment (see <http://moodle.org>). Moodle allows interactions to take place in a closed virtual environment, a 'safe community' that course organisers can monitor in case potentially sensitive topics cause offence or upset to the participants – and, if necessary, take remedial action. That said, while topics to date have occasionally touched on issues such as remembrance of war dead, sexuality and religious sectarianism, participants have handled these topics with sensitivity and mutual respect (Phipps and Corbett 2006).

The relative novelty of online communication has diminished over the last decade, with online environments such as 'moodle' now competing with social media such as Facebook. In a post-course interview one participant acknowledged awareness that online chat is governed by conventions, even if these remain implicit:

> I suppose I'm probably like part of the first generation which is like a computer native [...] cos myself and everyone that communicates using computers, there is a format for online communication, in so much as it's not codified and we don't discuss it, there is a way that people talk. (interview data)

The data suggest that there are indeed expectations that online communication will be characterised by its own order of indexicality, or 'way that people talk'. In the interactions these expectations were occasionally challenged, as when Glasgow-based participants were perplexed by the alternative conventions of Asian 'smileys'.

The participants

The main cohort of participants was a Glasgow-based group of largely British final-year undergraduates, with some European exchange students and occasionally some Asian postgraduates who audited. Each year they numbered around 20, but they were joined online by international participants, whose number and locations varied: in different years, they were based in Argentina, Brazil, Germany and Taiwan. The number of international participants generally outnumbered the Glasgow-based participants by 4:1; however, most of the topics were initiated by the Glasgow-based students, and much of the interaction was directed as much to the local as to the global audience. The exchanges usually have little explicitly to do with issues of global justice, and it is fair to say that the discussions covered in this chapter remain at a relatively mundane intellectual level: they are 'chat' rather than 'serious' discussion. However, we suggest that 'chat' can function as an index of intercultural communicative competence insofar as its main purpose is to display, negotiate and shape attitudes. Byram (1997: 34) asserts the importance of attitudes in intercultural learning:

> Attitudes which are the precondition for successful intercultural interaction need to be not simply positive, since even positive prejudice can hinder mutual understanding. They need to be attitudes of curiosity and openness, of readiness to suspend belief and judgement with respect to others' meanings, beliefs and behaviours. There also needs to be a willingness to suspend belief in one's own meanings and behaviours, and to analyse them from the viewpoint of the others with whom one is engaging.

The 'attitudes of curiosity and openness' that constitute a goal of intercultural communicative competence should, in theory, be facilitated by online interactions between participants from different countries. Certainly, the prioritising of attitude over content is evident in the following extracted sequence from a longer thread on 'fish and chips'. However, it captures the character of many of the postings; while breezy in character and usually undemanding in subject matter, these postings can be demanding in level of language and local cultural knowledge – here, for example, it is assumed that overseas participants will know about Whitby and Seahouses (two relatively small coastal towns on the Yorkshire and Northumbrian coasts, respectively). Two of the Argentinian responses confirm that the culinary institution that is taken for granted by the Glasgow-based participants only exists for them as part of the mythology of 'Britishness' drawn from literary representations, and they respond to the local scale of the original posting by offering an equivalent.[2]

> Hello!
>
> As we were preparing for one of our presentations, [firstname] and I discovered that people abroad no nothing of the "fish 'n' chip shop". This was something we were absolutely horrifed about, and we intend to make everyone aware of them!
>
> If anyone who has never had fish and chips from a fish 'n' chip shop GO RIGHT NOW AND GET ONE! You'll never regret it. I promise. You'll love them.
>
> If you're ever in Whitby or Seahouses, or any other seaside town, buy fish and chips there, if possible because they are the best! (Though make sure you're really hungry, because the fish tends to be HUGE!)
>
> Anyone who loves or hates (perish the thought) fish and chips, or if you've never heard of them and want to find out more about them, please share your thoughts! All comments welcome, good or bad!

> Hello
>
> You are right, I'm from Argentina and the only thing I know about "fish 'n' chip shops" is that they are named very often in British novels. We don't have that kind of shops here.

> Hello!
>
> I'm [firstname] from Argentina. It's true that here in Argentina we don't have that kind of shop. However, we do have our equivalent here. If you ever happen to be in Argentina, more precisely in La Plata, you must try our special: "milanesa con papas fritas". Here in La Plata, young people go to what can be called a shop (generally in a square or park), where the special is prepared at any time. Generally, after going to a disco, like at 5am in the morning, we have "breakfast" with one of this specials!!!!!!!!!!!!!
>
> No chippies in Argentina?!!!!!!!
>
> Well, if ever you see one anywhere in the world, you must try some fish and chips! Heaven!

2 Here and throughout, the spelling and presentation of the original extracts are retained.

In the 'virtual bonding' that occurs – or often fails to occur – in online intercultural chat, we may see, for example, the extent to which 'curiosity about the Other' is or is not made manifest in the small talk between young adults using computer-mediated communication to explore their own and other cultures. At one level, the British participants open up a discussion that prompts an interested response from the Argentinian e-partners. The exchange is friendly and open, and the emotional engagement is indicated, on both sides, by a super-abundance of exclamation marks, which may be indicative of the fact that the postings here are mainly by females (Waseleski 2006). In the terms of Ushioda (2011), the postings function as a means of expressing aspects of transportable identity (i.e. as lovers of fish and chips/*milanesa con papas fritas*). The Argentinian participants find a discursive space for themselves in a localised discussion first by relating the transportable identity expressed to a national stereotype (thus jumping scales to a translocal point of reference) and then finding an equivalent transportable identity to bring to the discussion as a point of comparison. But there is no further acknowledgement amongst the Glasgow-based participants about how their own transportable identity translates into a national stereotype, or how their own transportable identity might find equivalents elsewhere. As an occasion for a deeper expression of 'curiosity and open-mindedness' on the part of the Glasgow-based participants, then, this discussion counts as a missed opportunity.

Transportable identities

The online chats are characterised by the co-construction of 'transportable identities' through the discussion of superficially 'trivial' topics that allow an overt display of attitude and emotion. The more popular topics tend to be thin on substance, but instead designed to allow and elicit affective expression; for example, participants shared information and photos of their pets, discussed words they did not like, and talked about whether Sundays were pleasant or boring. Some topics took the form of lists of preferences, a constant feature of online communication, typified by 'What's your favourite drink (not including alcohol)?' and 'What's on your bucket list?' (i.e. the list of things you wish to do before you 'kick the bucket' or die). Some threads championed something that the initiator regarded as constitutive of personal or group identity, such as a football team or 'national' food. A very few threads raised questions about cultural conventions ('What gifts should you take when visiting people?', 'How do people in different countries commemorate their war dead?'), and changing social mores (attitudes to the introduction of non-smoking legislation in Scotland). All the discussion threads can be related to more or less substantial 'transportable identities', the latent social attitudes and stances that students may or may not evoke in the classroom.While the topics evoked online might seem superficial or trivial, they do have the virtue of representing the private aspects of the self that 'official' classroom discourse often overlooks (Ushioda 2011: 206). Interview participants commented, unprompted, on the nature of the majority of the topics proposed. Topics which involved sharing common interests were noted to be popular, as these were generally 'safe' topics with no right or wrong answer, and which, importantly for some, required little thought: "[...] when you talk about random things, the one thing comes and then the other. And the whole conversation went like crazy. So it was really nice." (interview data).

Some, revealing how they had looked back over previous exchanges in the run-up to the final exam, explained that what had seemed to be trivial topics at the time of posting actually held potentially deeper cultural significance which could be unpacked and reflected upon in light of subsequent postings and discussions. Similarly, several participants commented that even apparently superficial topics on a forum have a distinct advantage over more formal sources of cultural information, such as books and articles, as they come unmediated, ready for further exploration and interpretation:

> [...] you're interacting with a person, and you're seeing, if you view it as data, you see the data yourself, and you can draw your own conclusions. [...] Erm whereas when you're reading an article or a book, rarely does that come without some sort of judgement. (interview data)

Such topics may also cover ground which more formal sources of information do not: one participant described in her interview a discussion, which for her had been particularly memorable, of participants' views on the appropriateness of taking dogs into restaurants. Though not high on any agenda for social justice, this kind of discussion addresses and invokes individuals' stance on issues of normalcy and otherness.

Constructing identity by expressing affect

A selection of moodle threads has been extracted for detailed analysis in the sections below. They are taken from different years between 2004 and 2011. The criteria for selection is that the discussion threads should be student-initiated and have generated a minimum amount of discussion (defined here as seven responses). Together, these threads make up a small corpus of 28,728 lexical items, which we have analysed using WordSmith Tools (Scott 2008) to create simple word lists and concordances. We are arguing that a key means of expressing transportable identity is to raise a topic that invites an affective response; here, therefore, we consider the language used to express and elicit affect.

There are various ways of expressing affect, most obviously perhaps through positive and negative adjectives. The most frequently-used in our corpus, with numbers of occurrences listed in brackets, include:

Positive adjectives	*Negative adjectives*
good (68)	bad (25)
great (46)	weird (11)
best (27)	strange (8)
special (26)	horrible (7)
nice (24)	
happy (18)	
amazing (12)	
cute (12)	
beautiful (7)	

Relating the simple occurrence of these expressions to instances of positive and negative appraisal demands some caution, however, as can be seen in the following examples:

1. I think that the smoking ban is a **great** idea (in theory)
2. And I am not a **great** fan of insects

3. I don't really want to get into any not-**nice** territory…
4. Bye and have a **nice** Sunday!

At times a particular adjective will or will not be used to express appraisal directly: *a great idea* explicitly expresses value, while *not a great fan* does not; while the other two examples show that the sense can be negated in context. However, it is interesting to note that positive adjectives so outnumber negative ones, suggesting a predisposition in online communication to express even negative appraisal in positive terms (e.g. saying *not great* or *not-nice* rather than *badly* or *bad*). It is also suggestive for online intercultural exchanges that two of the more frequent 'negative' expressions express abnormality (*weird* and *strange*). In context, at least some of the usages refer to unfamiliar cultural routines or behaviours:

1. I was speaking to a French friend the other day who thought it was very **weird** that a lot of people in the UK go to the pub on Christmas Day.
2. It's so **strange** that dinner is eaten so late!

Again, in context it is clear that these expressions are not actually making negative appraisals, so much as indicating that the participant is reporting or experiencing a 'decentring' from a taken-for-granted set of assumptions about everyday life.

The most frequent verbs used in expressing appraisal in our corpus are, unsurprisingly, *like* (55 occurrences) and *love* (52 occurrences). These appear predominantly with first-person subjects, (40 for *like*; 43 for *love*); there are a few third-person reports of appraisal, and the remainder elicit a response from the correspondent. Examples are given below:

1. the thing I would **like** to do for the rest of my life
2. I also **like** dancing
3. I really don't **like**, actually I hate it
4. a dog lover myself…I do **like** cats, but…
5. we would **like** to share our bucket list
6. and generally I **love** animals. Once
7. have to agree with [firstname] that I **love** shopping! However
8. one of the hard ones and I **love** it when the bottom goes all squishy

9. swimming with dolphins…I would really **love** to do it!
10. I really **love** to visit quite different places

From these examples, we can see *like/love* express bald statements of appraisal, both positive and negative (*don't like*); desire (*would like/love*); and qualified approval (*I do like, but…*).

As noted, relatively few uses of the verbs are in the third person or in the second person. Examples of third-person uses are:

1. can understand why people don't **like** football with all the violence
2. anything anyone would **like** to know about Rangers
3. Many Dutch people **love** to go to French campings
4. I don't understand people who **love** snakes though

The first example is embedded in a display of empathy by a female football supporter for those who do not share her perspective; the second is part of the same participant's offer to share information about her favourite team. The third is a generalisation by a Dutch participant, and the fourth an acknowledgement that the participant does not have empathy with a particular group of people.

Perhaps surprisingly, elicitations of appraisal are relatively few. Examples include:

1. Do you **like** risotto? It's my favourite
2. Do you **like** desesperates housewives? [sic]

Here the participants seem to be expressing appraisal in order to identify and bond with people with shared tastes, rather than expressing curiosity about the interests of other participants, *per se*.

To summarise, in online intercultural exchanges in which participants initiate their own topics, the actual substance of their discussions may matter less than the opportunities they afford to construct a transportable identity through the expression of appraisal. While our data suggest that participants are adept at expressing their own opinions using the language of positive appraisal, they are relatively slow in valuing the responses of others, and in drawing their fellow participants out. When this does occur, however, it can lead to valuable exchanges of views and observations.

Negotiating conflicting identities

One of the more revealing discussion threads illustrates a complex negotiation of transportable identities in local and translocal contexts. The discussion centered on the 'transportable identity' of being a female football fan. Started by a Glasgow-based student from the north of England, the discussion was enthusiastically taken up by a participant from Glasgow itself, who declared herself a passionate Rangers supporter. Later in the discussion, a female Celtic supporter from Glasgow also joined the thread, and there were contributions from international as well as Glasgow-based correspondents.

The transportable identities of football fans are obvious occasions for either bonding or conflict, particularly in the local context of Glasgow. Rangers and Celtic, known jointly as 'The Old Firm', are the two most successful Scottish football clubs, and their long rivalry is coloured by their historical associations with the Protestant and Catholic communities in Glasgow, respectively. Enthusiasm for one club or the other can turn to expressions of tribal loyalties and, at worst, sectarian abuse and violence. On a broader level, Bradley (1995) argues, for example, that support for Celtic is a visible marker of an otherwise publicly effaced immigrant Irish identity in Scotland. Football fandom in Scotland is also associated with working-class males, though its traditional constituency has changed (Holt 1986), so the discussion of contemporary football by fans who are female undergraduate students is interesting from several perspectives. In particular, it is of interest to observe how female fans negotiate between the pleasure they derive from following their team and historical aspects that they find troubled and troubling. The discussion thread is extensive, so only edited extracts of a selection of the posts are considered here: this obviously colours the nature of the interaction somewhat. Four people are involved in this interaction: a self-proclaimed Newcastle fan (NF), a Rangers fan (RF), an FC Kaiserslauten fan (KF) and a Celtic fan (CF).

As well as Harry Potter and eating fish n chips, I also enjoy football as it's a full 90 minutes of suspence – anything can happen in that 90 minutes, for better or for worse.

I thought I'd open up a new topic on football, as I know that football is such a world wide phenomena, and some people just go mad over it!

I know hardly anything about foreign football, and until I moved to Glasgow, I knew next to nothing about Scottish football, which surprised me a lot, since I've enjoyed it since about 1996!

I support Newcastle United, and they're unfortunately third from bottom in the league due to our terrible performance this season! So, because I'm a newcastle fan (AKA a Geordie) I therefore am obliged to hail Alan Shearer as a football God and hate Sunderland AFC (AKA Mackems) – I can easily forgive Newcastle for losing to any team other than Sunderland!! And the feeling is mutual, I assure you!!

Though, there are also bad experiences from these football rivalries, as violence and verbal abuse often follows, especially if one team humiliates the other. Also, we're not often allowed to wear football strips in pubs in case it provokes some kind of violence later in the night.

I also support England (which is a dangerous thing to admit as I live in Scotland, so please don't hold it against me!!!!!) And I think our performance in the World Cup was poor. People say it was due to Rooney being injured. I say that Rooney is one man and can't win the world cup for England all on his own!!

[…]

For anyone who hates football … why? I can completely understand why people don't like football with all the violence and hooliganism that is associated with it.

Please, share your views, I would be extremely glad to hear them!

Initial post (NF: edited extract)

> YAY!!!!! I'm soooooooo happy that a football (fitba) forum has been started. Thank you [firstname], you have made my year.
>
> I absolutely LOVE football and have realised recently that I spend most of time talking about football. Although, recently it hasn't been too positive. I am a (Glasgow) Rangers fan…not just a fan…I am completely obsessed! I do love my team immensely but we are going through a really nasty spell at the moment which has caused a few tantrums, you know…kicking walls, punching floors, shouting, etc. and it has also caused some sleepless nights…I told you I am obsessed!
>
> […]
>
> Football colours aren't allowed in Glasgow pubs because of course in Glasgow we have a huge rivallry between 2 teams…Rangers (West End) and Celtic (East End). These 2 teams are commonly known in Scotland as "The Old Firm" and to be part of an "old Firm" team is the most incredible thing (although many players at Rangers at the moment don't seem to understand that, hence our really poor performances of late).
>
> […]
>
> I'd love to hear everyone's views on football as it is a major part of my life and a huge part of Glasgow culture and a really important part of our culture too. If there's anything anyone would like to know about Rangers, or Scottish football, don't hesitate to ask!

Response 1 (RF: edited extract)

> Hi all!
>
> I'm a football fan, too, but I cannot say I'm obsessed with footbal Anyway, as I'm from Germany, I support a German team of course: the 1.FC Kaiserslautern. The fans and the team are known as the "Red devils" and our stadium is logically "hell" for other teams. They are quite good at the moment… in the second league cause they weren't good at all last year and the stadium was more hell for the fans than for any other team.
>
> At the moment I don't watch the matches. But at home I often watch them on TV with my family, which is crazy about football. And once or twice a year I go to the stadium, normally with my parents as well.
>
> I was never able to watch a game quietly. I don't understand why anybody would do that. Football is such an emotional thing! At least I don't swear… I think…

> I wear a football shirt only if I go to the stadium, although it isn't a problem at all in my town to do so also in a pub. We only have one team and additionally the fans from my region are not known for violence of hooliganism at all. But a lot of them will get drunk to celebrate just like here

Response 2 (KF: extract)

> Hey [firstname]!
>
> I was just wondering about Rangers fans today … why is it that everywhere in Glasgow, I always see Celtic strips, like EVERYWHERE, and hardly any Rangers fans seem to wear their strips? I have only seen a few people wearing Rangers strips, especially in the city centre, I was just wondering why that was, since both sides seem to have huge fan bases!?
>
> I have even seen Celtic tops (but never Rangers) where I live in Hexham, and one or two in Newcastle (amid the many black and white strips!).
>
> Glad you're another enthusiast!! (And I'm glad I'm not the only girl who LOVES footy!!)

Response 3 (NF: extract)

> That was my exact thought yesterday…a girl who loves footy…I'm not the only one!! YAY!!!
>
> My answer to your question would be…I don't really know! My funny and Rangers fan answer would be that Celtic fans are glued into their strips because they are frightened to take them off. Maybe they feel like less of a fan if they wear something normal for a day. No seriously…no offence to any Celtic fans out there.
>
> […]
>
> I should apologise to anyone reading this who is a Celtic fan. It is very difficult to put your points across regarding football, particulary if you are from Glasgow and a supporter of the Old Firm. It is easy to be bias at times like these and I apologise if I have come across this way! I would love to hear from Celtic fans for their opinion on the matter.
>
> Again [firstname], thank-you so much for starting this forum cos talking about football is my favourite thing to do!

Response 4 (RF: edited extract)

> Hi everyone!
>
> I have been waiting a while to reply to this forum as there is lots of things I want to talk about – in fact, probably too many things to talk about! As you probably know from previous posts I am a Celtic fan from Glasgow. I can't say that I am obsessed with football or Celtic, but, I really enjoy watching and talking about it and think I know enough to contribute to this!
>
> I don't think nowadays that football is just interesting to boys/men; women enjoy the game for several reasons! I personally became interested because of my Dad and Brother's love of Celtic, and then my fiance came along and sucked me into the celtic way even more! So, now I can say I am officially a loyal and happy GHIRL!
>
> […]
>
> I don't really want to get into any not-nice territory but violence is obviously something all football fans are aware of; everyone should just keep to themselves. Sure, support your team and enjoy the good times, but, if something does go wrong, accept it and behave properly! I feel very intimidated sometimes in work; songs are sung without even a tiny thought for me or other customers with different views/religions. We have friends that come from a particular area of Glasgow filled with Rangers fans – my fiance can't wear his Celtic top when visiting for fear of his life! It is bad at the moment, but, hopefully things can get better!…
>
> I have some very close friends that are Rangers fans and I am all for hearing about Rangers and their successes – as long as you don't mind me talking about Celtic too! It's all good when you can have a laugh about things!
>
> I know I have gone on but I'll finish now by saying that I love the Celtic strip and believe that Celtic fans wear their strips because they are so proud of their team and all that surrounds them! I think the strips are attractive and CERTAINLY NOT GARISH MISS [surname]!!! My dad doesn't like to see girls in football strips so I only got my first one when I was sixteen! I love to see Celtic strips abroad, or any Scottish teams for that matter – it makes you feel at home!
>
> Speak soon and happy chatting!
>
> Luv'n'hugs, [firstname] xxxxx

Response 5 (CF: edited extract)

At one level, this interaction can be compared to the 'fish and chips' thread discussed above. There is a tension between the scales at which the discourse develops: the local cultural references in the postings may inhibit contributions from translocal participants. We see that in order to participate, KF has to find her own equivalent to the Glasgow rivalry and to relate her own attitudes and behaviour as a fan and spectator to the localised behaviour and attitudes evoked in the construction of NF and RF's transportable identities. As in the 'fish and chips' thread, the Glasgow-based participants fail to respond in much detail to the translocal perspective that KF provides.

Even so, all participants recognise the ethical issues involved in 'loving' one team by 'hating' its rivals, especially the local rival, and yet they endorse the 'hatred' while dissociating themselves from violence, verbal abuse, drunkenness or overtly insensitive displays of sectarian or other chauvinism. NF distances herself from 'authentic' hatred of Sunderland fans by recognising that it is an 'obligation' and acknowledging her potential position as an object of hatred as an England fan in Scotland with the plea 'please don't hold it against me!!!!'. In most of the postings the liberal use of exclamation marks and capitalisation is again consistent with findings that suggest that females use a greater proportion of explicit exclamations than males to express a range of emotional states, such as excitability, supportiveness and friendliness (Waseleski 2006). KF uses the fewest exclamations, the one instance in the sample being, appropriately, 'Football is such an emotional thing!' The effusive exclamations counter possible interpretations of the postings as antagonistic.

The postings nevertheless acknowledge that they are treading a fine line between supportiveness, banter and abuse. In response to NF's question about the ubiquity of Celtic strips, RF gives a 'funny' answer denigrating Celtic fans, then apologises for giving possible offence, explicitly raising the difficulty of not displaying 'bias' and invites Celtic fans to contribute to the thread. She reinforces her supportiveness by thanking NF for starting the thread and reiterates her own enjoyment of the discussion.

Together, the postings show the participants carefully negotiating a topic (perhaps epitomised by CF's measured and careful contribution) where the obligation to 'hate' the other is mitigated by mutual respect. The participants do this by explicitly distancing themselves from acts of violent and abusive chauvinism, while asserting their pride in their own team's achievements and iconography. Their transportable identities are situated in a context of domestic, local and national loyalties, and in the context of these loyalties the participants accept that their identity is in part defined by rival fans' 'hatred'. To avoid 'authentic' antagonism, they display supportiveness through acts of thanking, pre-emptive apologies, friendly salutations and the effusive use of exclamation marks. While the postings show an awareness of sectarian violence, there is, perhaps understandably, little depth in the discussion of this issue; indeed CF voices a preference for avoiding such discussion. Reflection on what it means to be a female supporter of a male-dominated game is also largely restricted to a discussion, later in the thread, of the difficulty of buying football strips that flatter the female figure, a topic on which the female fans of rival teams are all able to find common ground.

Territories and interventions

A pedagogical issue that is raised by the online intercultural exchanges, and explored in the post-course interviews, is the degree to which teachers should intervene in students' evolving dialogues. As course facilitators, the lecturers made the conscious decision to stay out of student forums as much as possible, since we felt that participants should largely be free to initiate and develop discussions as they wished. This decision had positive and negative outcomes: when the dynamic between participants was good, some lively discussions arose that touched on areas that the lecturers would have had difficulty in predicting. In some years, however, the participants failed to initiate discussions that provoked sustained responses; and even when discussions were sustained, as in the 'football' thread, it might be argued that lecturers could have prompted a greater depth of reflection on the aspects of identity and belonging that were being raised, almost incidentally, as the thread developed.

Scaffolding

Occasionally a lecturer did intervene in the discussion to attempt to provoke a greater degree of reflection. In the thread on 'fish and chips', which the initiators mock-seriously presented as a campaign to raise international participants' awareness of this national dish, the course lecturer became involved in the exchange:

> Okay – I can't resist the fish 'n' chips campaign.
>
> So – why do you all think fish and chips are so popular in the U.K.? And why is a fish and chips campaign important for intercultural communication?
>
> Just wondering .....

Lecturer intervention

> Hmm...
>
> Well, we said in the lecture that since we are an island we are arguably surrounded by fish and we have had a great fishing industry and tradition for centuries, and potatoes grow very well here, so we have the ingredients pretty much sorted.
>
> However, since potatoes are from S America – a link to our friends in Rio!! – chips are technically French, and according to one Professor Panikos Panayi we've pinched ideas from Jewish fish dishes, perhaps this dish is more intercultural than we like to imagine! I think this is partly why this campaign is important for intercultural communication: even if it has roots elsewhere, when something is so important for and internationally attached to a particular nation there is always some great tradition or story behind it and, from what I understand, traditions and social identity are what 'culture' is built on.

Response to lecturer's intervention (extract)

Here the lecturer's intervention did prompt a Glasgow-based student to do some further thinking and research on the multicultural origins of the 'national dish' and to come to some initial understanding of the heterogeneous and discursive construction of 'culture'. The participant's response maintains the sense of involvement with the peer community ('a link to our friends in Rio!') while presenting a set of statements whose tentative nature is indicated by hedging features (*arguably, perhaps, I think*, etc). The intervention here, coming after a series of relatively unreflective postings, can be judged a success; at other times, however, lecturers' interventions had the effect of bringing a discussion to a halt.

The question of the extent to which lecturers should become involved in participants' online discussions was one explored in the post-course interviews. Opinions were surprisingly mixed: while many appreciated the freedom and space to raise topics of their own, others commented that they would have liked more direction, encouragement and, for some, even a requirement to post messages:

> I thought it was quite good that there were no, say that you and John wouldn't comment on anything that was said or that you would try to have the students discuss amongst themselves, because otherwise, at some point, people would just be like, 'Okay, so somebody wrote something, so let's see what John or Wendy says, and then we'll say something.' (interview data)

> I think if you tell students they should be more active, and make them do it more, then they will get involved […] maybe if tutors had more of a role then they would see, 'okay we should be doing this, because they're doing it, so we must be doing it as well.' (interview data)

> it's something that I've noticed elsewhere where there have been sort of online learning forums, erm, where it's part of the course content, explicitly, people are generally only, or I think people will only take part in it when they've been explicitly told 'Do that'. (interview data)

The status of the lecturers also prompted different views: some were very comfortable with the fact that tutors could be just as unfamiliar as the students with certain cultural attitudes and practices, while others found it difficult to move beyond a traditional view of a tutor as someone with a certain level of knowledge and authority, with one commenting, at the end of the course, that 'you don't really see your teachers as part of the group'.

Conclusions

The topics discussed here do not directly evoke the mainstream issues of 'global justice' that one might expect of a contribution to this volume. Nevertheless, we believe that the analysis of online 'chat' merits a place in these discussions for several reasons. First, the 'transportable identities' constructed by online chat represent a set of social identities often neglected in language classrooms. In suggesting a 'dogme' approach to language learning, Thornbury (2000) puts 'learner talk' at the centre of the curriculum, and Ushioda (2011) suggests that online communication presents an opportunity to focus on issues generated by learners themselves, and to 'scaffold' their articulation of attitudes and identities. Here we focus on learner-generated conversation rather than teacher-imposed agendas. One issue, however, is the nature of the teacher intervention and how and when 'scaffolding' is appropriate.

Generally, as O'Dowd (2012: 353) reminds us, the development of intercultural communicative competence takes place in the wider educational context in which online exchanges occur:

> Studies have shown repeatedly that learners need the guidance and informed insight of their teachers to create their own online correspondence and to interpret and respond to the messages, blog posts and video recordings that they receive from their partners. It is in the classroom analysis of these authentic foreign texts that cultural 'rich points' emerge and the skills of intercultural interaction and ethnographic interviewing can be honed.

To this we would only add that the 'authentic texts' need not be entirely foreign: the postings above show that students have much to learn from the analysis of language and interpersonal discursive strategies used in their own online postings. It may be that the classroom analysis and reflection which is required here means that it is easier to appreciate the development of intercultural communicative competence when reflecting on it after the event. Certainly, most participants found themselves able to identify aspects of their own learning by the time of the post-course interviews, which took place after the final exam for the course. Some took very practical things from the course, such as tips on how to approach culture in a possible future job as a language teacher, or how to structure a language learning task. Others, like those quoted below from our interview data, focused on less tangible learning that is nevertheless clearly linked to their understanding of their own and others' identities:

> I'm probably a lot more aware of cultural behaviours and cultural differences […]. Erm, and maybe more curious than I was to explore those differences.

> But I think you kind of learn just to kind of, you don't, I don't think you sacrifice your own beliefs, but you kind of look at it from a, you can see it in a more neutral way. If you see what I mean.

Overall, one of many lessons we have learned from the project is to take 'chat' seriously as both a problematic and revealing phenomenon in translocal, intercultural communication.

References

Anderson, W. and Corbett, J. (2012) Shaping intercultural competence? Creating a virtual space for the development of intercultural communicative competence. In: Jamarani, M. and Sharifian, F. (eds.), *Language and intercultural communication in the new era*, pp. 99–115, New York: Routledge.

Belz, J. A. (2003) Linguistic perspectives on the development of intercultural competence in telecollaboration. *Language Learning & Technology*, 7(2): 68–99.

Belz, J. A. (2007) The development of intercultural communicative competence in telecollaborative partnerships. In: O'Dowd, R. (ed.), pp. 127–166.

Blommaert, J. (2010) *The sociolinguistics of globalization.* Cambridge: Cambridge University Press.

Bradley, J. M. (1995) Football in Scotland: a history of political and ethnic identity. *The International Journal of the History of Sport*, 12(1): 81–98.

Byram, M. (1997) *Teaching and assessing intercultural communicative competence*. Clevedon: Multilingual Matters.

Byram, M. (2008) *From foreign language education to education for intercultural citizenship: Essays and reflections.* Bristol: Multilingual Matters.

Corbett, J. (2003) *An intercultural approach to English Language Teaching.* Clevedon: Multilingual Matters.

Corbett, J. (2010) *Intercultural language activities. Cambridge Handbooks for Language Teachers.* Cambridge: Cambridge University Press.

Corbett, J. and Phipps, A. (2006) Culture, language and technological control: virtual intercultural connections. In: Linke, G. (ed.) *New media, new teaching options?!* pp. 157–178, Heidelberg: Universitätsverlag Winter.

Holt, R. (1986) Working class football and the city: the problem of continuity. *The International Journal of the History of Sport*, 3(1): 5–17.

Martin, J. R. (2000) Beyond exchange: APPRAISAL systems in English. In: Hunston, S. and Thompson, G. (eds.) *Evaluation in text: Authorial stance and the construction of discourse*, pp. 142–75, Oxford: Oxford University Press.

Meskill, C. and Anthony, N. (2010) *Teaching languages online*. Bristol: Multilingual Matters.

O'Dowd, R. (ed.) (2007) *Online intercultural exchange: An introduction for foreign language teachers*. Clevedon: Multilingual Matters.

O'Dowd, R. (2012) Intercultural communicative competence through telecollaboration. In: Jackson, J. (ed.) *The Routledge Handbook of Language and Intercultural Communication*, pp. 340–356, Abingdon: Routledge.

Richards, K. (2006) 'Being the teacher': Identity and classroom conversation. *Applied Linguistics*, 27(1): 51–77.

Risager, K. (2007) *Language and culture pedagogy: From a national to a transnational paradigm*. Bristol: Multilingual Matters.

Scott, M. (2008) WordSmith Tools, version 5. Liverpool: Lexical Analysis Software.

Thornbury, S. (2000) A dogma for EFL. *IATEFL Issues,* 153 (Feb-March), p. 2.

Ushioda, E. (2011) Language learning motivation, self and identity: current theoretical perspectives. *Computer Assisted Language Learning*, 24(3): 199–210.

Waseleski, C. (2006) Gender and the use of exclamation points in computer-mediated communication: An analysis of exclamations posted to two electronic discussion lists. *Journal of Computer-Mediated Communication*, 11(4): 1012–1024.

Zimmerman, D. H. (1998) Discoursal identities and social identities. In: Antaki, C. and Widdiecombe, S. (eds.) *Identities in talk*, pp. 87–106, London: Sage.

Marina Orsini-Jones, Elwyn Lloyd, Zoe Gazeley,
Gwenola Bescond and Beatriz Vera López

Student-driven intercultural awareness raising with MexCo: agency, autonomy and threshold concepts in a telecollaborative project between the UK and Mexico

This chapter reports on an international intercultural telecollaborative project – MexCo – involving students and staff at Coventry University UK (Department of English and Languages), staff at the University of Warwick UK (Language Centre) and staff and students at the Universidad Nacional Autónoma de México (Departamento de Lenguas Extranjeras, Facultad de Estudios Superiores Zaragoza).

The aim of MexCo was originally to investigate online tandem language learning, but it subsequently developed into a wider study aimed at investigating its participants' "cultural frames" as defined by Gumperz and Roberts (1991). Students designed digital learning objects based on their telecollaborative forum postings that provided revealing insights into their perceptions of what constitutes 'culture'. Data collected demonstrate that the project has given students the opportunity "to confront and deal with the prejudices, stereotypes and myths that they hold about other social groups and cultures and that others may hold about them" (O'Dowd 2007: 29).

Areas of troublesome knowledge emerged.The linguistic exchanges in the dedicated asynchronous *Moodle* forum and the analysis of the transcription of the semi-structured interviews, brought to the fore the ways in which language reveals perceptions of power relationships amongst the agents involved, as discussed in critical discourse analysis (Fairclough 1995). The nature of the socio-cultural constructions emerging from MexCo and the way many students were able to review their initial perception of 'the other' during their transformative telecollaboration journey highlighted the power that such projects can exert towards the support of a raised awareness of cultural difference. At the same time, both staff and students involved had to confront some emerging "troublesome knowledge" in the exchanges that related to cultural hegemony as discussed by

Gramsci (1947). The perception of English as a language of power appeared to be the "tacit knowledge" (Polanyi 1958) shared by participants. The data is therefore yielding interesting insights into the connections between language, power and culture both in terms of specific issues relating to the pedagogy of second language teaching and learning and wider concerns relating to educational social justice. This relates to Fraser's (1996) "politics of recognition" that should be embedded into the higher education curriculum to ensure that graduates are ready to operate in a "difference-friendly world" (p.3). It also reinforces Kumaravadivelu's (2001) stress on the need to embrace a postmodern approach to language teacher education and explore the particular socio-cultural setting of all agents involved in the learning experience to make it relevant and fair to them. Kumaravadivelu's postmodern modelling of ELT teacher education links with a wider movement in sociology from "solid modernity" to "liquid modernity", where "the prime concern…needs to be the promotion of autonomy and freedom" and the need for a raised individual self-awareness and sense of responsibility for all agents concerned (Bauman 2000: 213).

Telecollaboration can promote autonomous learning and generate ways of creating new content and learning contexts which are both personalized and collaborative. One of the aims of the MexCo project is to design culturally sensitive ELT online materials that have been informed by the Mexican students' interests in a 'bottom up' way. ELT materials where English is changed from within by the local needs of its learners who directly impact on it as agents of change; a change that is enabled by a telecollaborative environment informed by postmodern principles.

Empowering Critical Multiliteracies and Challenging Multimodal Digital Competences

The affordances of the e-tools used in MexCo played an important part in the project, and the analysis of the online exchanges also highlighted issues relating to "cyberpragmatic competence" (Yus Ramos 2011) – skill in understanding others' intended meanings in computer-mediated communication. As discussed in previous work (Orsini-Jones 2010), the acquisition of digital multiliteracies and the ability to move across different genres (digital and not digital) can be empowering for students. Freire and Macedo (1987), quoted in Hokstad and Dons (2007), maintain that literacy is "the ability, the possibility and will to read the world" (p. 2). Warschauer and Ware (2008: 228) quoting Castells (2000) stress that digital competences can enable all learners to be "interacting" rather than be passively "interacted". E-tools can and should be used in order to promote learner autonomy and the crossing of conceptual and cultural thresholds that bring about changes in the identity and agency of the learner.

Edward Snowden's whistle-blowing "affair" reporting on how the USA was spying on its closest EU allies using state-of-the-art cyber-bugging technology has revealed one of the darkest sides of what intelligence services can do in a connected world (MacAskill and Borger 2013). But this story has also illustrated that very well established power-structures can be undermined by viral dissemination via Web 2.0 tools which were in part instrumental in the creation of said structures. New technologies are therefore a double-edged sword and should be integrated into the higher educational curriculum taking account of both ethical frameworks and the addressing of possible inequalities. The advent of Web 2.0 technology is making the development of critical digital literacies more urgent because students, now exposed to a multitude of new information, must quickly take decisions regarding the value and validity of this information. Siemens (2004) quoted in Mason and Rennie (2008: 19) argues that:

> Decision-making is itself a learning process. Choosing what to learn and the meaning of incoming information is seen through the lens of a shifting reality. While there is a right answer now, it may be wrong tomorrow due to the alterations in the information climate affecting the decision.

A positive aspect of the opportunities and affordances offered by Web 2.0 tools is that they can bring about a democratization of the learning environment, where a tutor cannot any longer be the 'sage on the stage' and needs to learn how to negotiate new knowledge collaboratively with learners (e.g. Warschauer and Ware 2008; Guth and Helm 2010, Helm and Guth 2010). O'Dowd (2013a) proposes that the Web 2.0 tutor should have specific competences in the following spheres: organizational, pedagogical, ICT/digital and in the area of interculturality (attitudes and beliefs). Many of the latter relate to the respect of "the other" in a globalized context, e.g. "an openness to partner teachers' alternative pedagogical beliefs and aims" (O'Dowd 2013:8).

During a plenary discussion led by O'Dowd (2013b) in an online conference held jointly by Coventry University and the University of Warwick in June 2013, whereby experienced telecollaboration tutors from all over the globe spoke of the major challenges they had encountered in online learning exchanges, the above competence was identified as one of the most troublesome. All participants agreed that intercultural international collaboration requires a humbling re-examination of one's beliefs and a considerable amount of patience and negotiating skills.

A further "postmodern" consideration is that all participants engaged in global telecollaborative exchanges experience a hypertextual dynamism in the sphere of learning as it is generally agreed that "by destabilizing and seemingly 'disordering' the academic text, digital media enable new forms of academic discourse, literacy and knowing to emerge" (Land 2011: 61). Parallels have been drawn between the new landscape of Web 2.0 educational possibilities and the way the concept of culture is being reinterpreted (Thorne 2010). This echoes what Guth and Helm (2010 and 2011) write about telecollaboration and the use of the Internet in language learning and teaching. They report that the Internet has both contributed to "the hegemony of English as a *lingua franca*" – as argued by Crystal (1997) and Sharifian (2009) – as well as "giving a voice to the diversity of languages in the world" – as argued by Warschauer (2002) and Hafez (2007) – (Guth and Helm 2010: 14). At the same time EFL (English as a Foreign Language) learners have been empowered by the realisation that they can communicate with others who have also learned English as a foreign language, including those with whom they share political views about the hegemonic role played by countries such as the USA and the UK (Ryan 2002). Kumaravadivelu (2012) argues that "cultures are in closer contact now than ever before, and are influencing each other in complex and complicated ways" (p. 4).

There is a further aspect of telecollaboration that directly relates to social justice: that of accessibility. Research shows that the Computer Mediated Communication (CMC) opportunities offered by Web 2.0 tools provide the more introverted learners less threatening socio-collaborative platforms than face-to-face ones (Lloyd 2012) and enable them to thrive. Moreover, educational technology can provide access to virtual mobility and international intercultural exchanges for students who either cannot afford to travel or are unable to embrace physical mobility. Two of the UK participants in the project MexCo did not have a passport, for example, and this was the first opportunity of 'international mobility' they had experienced (Orsini-Jones, Lloyd, Gazeley, Bescond and Vera López 2013). Telecollaboration can therefore remove barriers to education and provide disabled students with learning opportunities that they would be otherwise unable to access (Hutchinson, Atkinson, and Orpwood 1998; Shakespeare 2007; Thomas and Loxley 2007).

Online international learning can moreover positively impact on the Higher Education (HE) curricula. In this confusing globalised context it becomes even more crucial that the teaching profession be aware of the fact that both the ability to engage in intercultural dialogue and acquire *academic* (not generic) digital multimodal multiliteracies are competences for global citizenship that students in HE must acquire. The importance of the acquisition of intercultural competence in general and Intercultural Communicative Competence (ICC) in particular (for language learners) has been stressed by a variety of HE-related bodies, both in the UK (Byram, Gribkova and Starkey 2002; HEA 2013) and in other countries (MLA Ad Hoc Committee on Foreign Languages 2007). It is nowadays generally agreed that a 'global graduate' should be able to both recognize and value cultural difference. The MexCo project discussed here provides evidence that telecollaboration can support this.

In line with threshold concept pedagogy (Meyer and Land 2003, further details below), we posit that interculturality is however troublesome and could be seen as a liminal space where the learner's identity is negotiated and reconfigured. The journey into another culture and reality ends as a journey in the rediscovery of one's own culture and reality and often in a healthy – but at times challenging – re-discussion of one's intercultural frameworks.

Intercultural Communicative Competence (ICC): Byram's ICC model and new Web 2.0 ICC considerations

Byram (1997) introduced the concept of the 'tourist' and the 'sojourner' in an attempt to explore how an individual relates to or communicates with a person in different cultural contexts. The "tourist hopes that their own way of living will be enriched but not fundamentally changed by the experience of seeing others", whilst the "sojourner produces effects on a society which challenge its unquestioned and unconscious beliefs, behaviours and meanings, and whose own beliefs, behaviours and meanings are in turn challenged and expected to change" (Byram 1997: 1). Byram (1997) labelled the quality that the "sojourner" required as "intercultural communicative competence" (ICC) and proposed a "schema of the factors involved – in ICC – and the relationship among them" (p. 33). Byram's key components of ICC are:

- Knowledge (savoirs): of self and other; of interaction: individual and societal
- Intercultural attitudes (savoir être): relativising self, valuing other
- Skills of interpreting and relating (savoir comprendre)
- Skills of discovery and interaction (savoir apprendre/faire)
- Critical cultural awareness and political education (savoir s'engager)

Byram (1997) believes that for learners to become competent intercultural communicators they need to be willing to suspend judgement on others' beliefs and behaviours as well as their own and that they should be open to analysing their beliefs and behaviours from the viewpoint of the person with whom they are engaging. This could be achieved through reflective activities.

As stressed by Helm and Guth (2010), Byram's ICC competences were drawn up and defined in the 1990s, they were mainly Eurocentric, and they presupposed a physical type of mobility, mainly involving students on Erasmus programmes, travelling to and studying in partner universities in Europe. The five "savoirs" were originally only applied within what Kumaravadivelu (2008) calls a 'modern' conceptualisation of culture, steeped in national identity, linked to clearly defined geographical boundaries. The affordances of Web 2.0 tools and the global arena that is now available for telecollaboration and online international learning have brought about a shift towards agency in a postmodern world; as argued by Kumaravadivelu (2008), "identity formation is conditional not merely by inherited traditions such as culture, or by external factors such as history, or by ideological constructs, such as power, but also by the individual's ability and willingness to exercise agency and to make independent decisions" (p. 144). Drawing from Barnett (1997) Vera López (2012) posits that the development of personal epistemologies and their integration with social epistemologies should be a priority for HE institutions. Helm and Guth (2010), O'Dowd (2007), O'Dowd and Ware (2009) and Thorne (2010) have discussed how the globalization of CMC and new digital affordances are altering the way in which we define ICC. The table provided by Helm and Guth (2010: 74) as an alternative to Byram's highlights that telecollaboration 2.0 includes as part of its goals not just ICC, but also new online literacies and an 'amplified' set of linguistic competences.

Threshold Concepts and Action Research

The identification of troublesome intercultural and metacognitive issues in language learning positions this study within the relatively new field of transactional educational inquiry known as "threshold concept pedagogy" (Cousin 2009: 209–212, Figure 1). According to Meyer and Land (2003, 2005, 2006; Flanagan 2011), a threshold concept is:

- Troublesome: the learners will often find it problematic;
- Transformative: once understood, its potential effect on student learning and behaviour is to occasion a significant shift in the perception of a subject;
- Integrative: it exposes the previously hidden interrelatedness of concepts that were not previously seen as linked;
- Irreversible: the change of perspective occasioned by acquisition of a threshold concept is unlikely to be forgotten.
- Bounded: a threshold concept will probably delineate a particular conceptual space, serving a specific and limited purpose.

Perkins (2006) argues that a threshold concept forces learners to question their assumptions, to reconfigure their learning landscape. A threshold concept can be 'counterintuitive' and cause the learner to feel displaced. Cousin (2009: 209–211) describes threshold concept pedagogy as a partnership amongst educationalists, tutors and students who 'tackle' together the scene of difficulty occasioned by the encounter with troublesome knowledge. The role-reversal model presented below by Orsini-Jones (Orsini-Jones 2010 – Figure 2) is dialogic and, while stemming from Cousin's, in it the identification of threshold concepts is driven by student researchers, who having embraced threshold concept pedagogy for their own research, help staff with identifying nodes of troublesome knowledge and help tutors to approach such nodes from the students' perspective.

This in turn causes tutors to transform their own thoughts and identity in pedagogical terms, as they can then observe their curricular intervention "through the looking glass" – to borrow Lewis Carroll's words (1871) – of the expert student's perspective (e.g Orsini-Jones 2013; Orsini-Jones, Brick and Pibworth 2013). The inquiry becomes therefore student-driven, rather than student-centred or tutor-driven. The "expert student" involved in the first phase of this MexCo project was Zoe Gazeley, who has co-authored this work (Gazeley 2013).

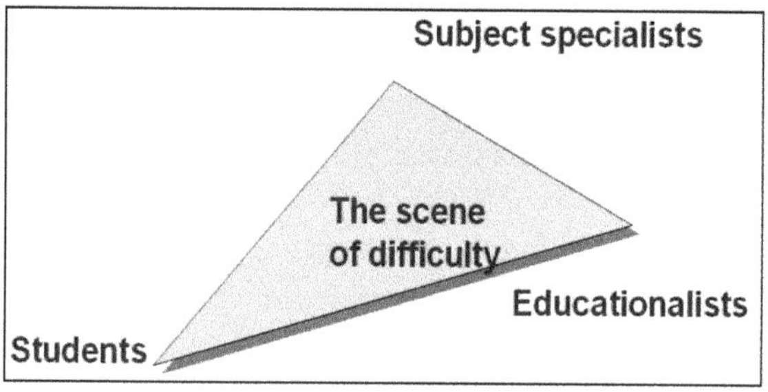

Figure 1: Threshold concept transactional model (Cousin 2009).

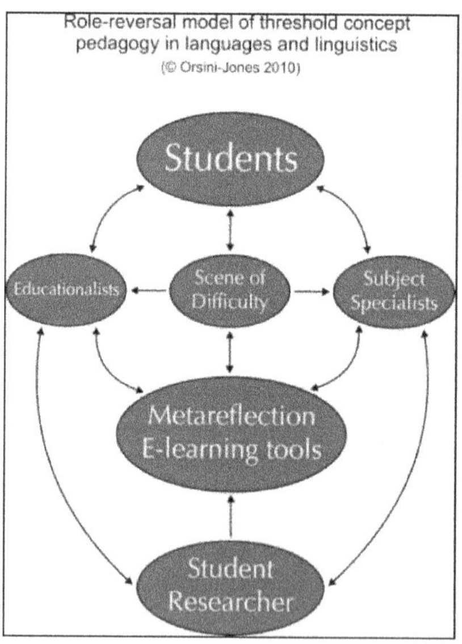

Figure 2: Role-reversal model of threshold concept pedagogy in languages and linguistics (Orsini-Jones 2010).

Land, Meyer and Smith (2008: ix-xxi) suggest that the identification of threshold concepts allows tutors to put in place targeted curricular interventions aiming at supporting students' learning experience and making such 'troublesome knowledge' less troublesome.

Issues relating to language and 'languaging' via subject specific epistemes are central to this field of inquiry. Meyer and Land (2005) stress that at the heart of threshold concept pedagogy is the "inter-relatedness of the learner's identity with thinking and language. Threshold concepts lead not only to transformed thought but to a transfiguration of identity and adoption of an extended discourse" (p. 375).

The research questions explored in the MexCo project in light of threshold concept pedagogy are the following:

- Is ICC a threshold concept?
- Does the engagement with digital multilingual multimodal multiliteracies help language learners cross the transformative portal leading them to "making sense" of what ICC is?
- What troublesome cyberpragmatics (Yus Ramos 2011) knowledge do language learners encounter in CMC global exchanges?
- Does online international learning help with overcoming stereotypical perceptions of the culture of others and scaffold an ethical reconfiguration of one's intercultural framework for all participants involved (staff and students)?

The above questions are being explored via participatory action-research cycles (McNiff 1988; Kemmis and MacTaggart 2005), where students actively inform each cycle in discussions with tutors.

The MexCo Project

In this section we explore how the themes and concepts discussed above informed the development and framing of the MexCo project, a telecollaboration/CMC knowledge-transfer intercultural project that started in April 2012, following informal contact between Vera López and Orsini-Jones, due to a shared interest for threshold concept pedagogy applied to the context of language learning. The project subsequently received funding from the Higher Education Academy (HEA, UK) and this enabled the initiators to also involve an expert on telecollaboration from the University of Warwick and set up project teams in each of the two countries involved. The UK cohort of undergraduate year 1 students who took part in it were from many different nationalities and backgrounds and included Romanian, Somali and Ghanaian students and many students of Pakistani origin (one hundred and fifty participated in total in the first two phases of the project). Most of them were not from Coventry. Their age varied between eighteen and forty five. They were studying on a variety of Languages and Linguistics-related degree courses (Single Honours and Joint Degrees in the subject areas of Teaching English as a Foreign Language (TEFL), Spanish, French, English Literature, English Linguistics and Creative Writing). The Mexican students were not all from the first year of their course; they were all from Mexico City or its metropolitan area and were studying English as a Foreign Language through their university-wide language provision. The age range was similar to that in Coventry. Unlike the Coventry University students they were not all studying humanities or social science-related subjects. Fifty-eight took part in the first two phases of the project.

The MexCo project's main aims are to:

- Develop intercultural awareness and ICC
- Develop multimodal digital multiliteracies
- Develop global citizenship

The project positioning fits within the following description by O'Dowd (2010):

> a third 'generation' or model of telecollaborative exchanges is emerging, which reflects in many ways a more flexible and adaptable interpretation of how online intercultural interaction and exchange can take place in foreign language learning contexts. […] this model focuses not only on the development of learners' linguistic and intercultural competence but also on developing the electronic literacies necessary to learn and work in today's information society. This emerging approach to online exchange allows for a much greater spectrum of possible partners, language setups and forms of interaction. (p. 3)

Unlike telecollaborative projects carried out and researched by others, e.g. the USA/Germany project by Schenker (2013), whereby the tutor prepared highly structured online environments and tasks, the approach in MexCo was deliberately 'bottom up' to begin with, student-centred and student-driven. Another distinguishing feature is that the project is informed by the relatively new field on transactional educational inquiry known as "threshold concepts" (Meyer and Land 2003, 2006; Cousin 2009) used in conjunction with action-research, as discussed previously.

The curricular integration of the online learning project is moreover underpinned by the following principles relating to e-learning design and research in language learning and e-learning:

- Long-term language projects (tasks) that involve student-centred problem-solving elements have been reported to have a positive impact on language learning (Ellis 2003; Klapper 2006) and enhance multilingual multiliteracies (Warschauer and Grimes 2007).
- An e-learning activity must be very carefully designed and is defined as a "specific interaction of learner(s) with other(s) using specific tools and resources, orientated towards specific outcomes" (Beetham 2007: 28).
- As implied by McLuhan (1967) the medium (or media) chosen for the task affects the students' learning experience and cognitive journey.
- Learning, as argued by Vygotsky (1962), "is a socially mediated activity in the first instance, with concepts and skills being internalized only after they have been mastered in a collaborative context" (cited in Beetham 2007: 36).
- Metacognitive activities, supported by e-learning tools, must be integrated into language learning tasks: thinking on how one learns can help with learning (Moon 2004).

In the pilot phase, which was relatively informal, students explored common themes of interest throughout the summer months of 2012 in an asynchronous discussion forum set up for this purpose in *Moodle*. They also communicated via *Skype*, email and *Facebook*.

ICC tasks based on the data that emerged from the informal pilot phase were discussed, agreed and proposed to the student participants in trilingual format (Spanish/English/French) to reflect the languages the students were studying, but students had the choice to design/choose their own tasks, and could post responses and comments in whichever language(s) they wanted to use in the forum.

With reference to the enrolment of participants on the dedicated website, some technical issues were encountered in terms of the effective registration of the Mexican participants in the first phase of the project. A new

VLE area was created towards the end of September 2012 where participants could easily self-register. Bilingual instructions (English/Spanish) were created for this purpose by staff at Coventry University and checked for accuracy by staff at FESZ/UNAM. An interesting aspect of this translating activity was the realisation by all parties concerned of how different Mexican Spanish was from Castilian Spanish – this topic will be explored further in a future study.

When Semester 1 started at CU in October 2012, the CU MexCo team linked the exchange project to a mandatory academic skills module on the English and Languages courses: *Introduction to Studying English and Languages at University* (worth ten of of one hundred and twenty credits in the first year[1]) while staff at FESZ linked it to their module *Inglés Curso General (CEFR A1 plus).* It was agreed to offer both CU and FESZ students the opportunity to engage with their peers in order to share both cultural and linguistic knowledge in keeping with the project's aim to enhance both the students' intercultural awareness and their employability capabilities and competences as 'global graduates'. This resulted in the collaborative design of assessed learning objects whereby students from Coventry and from Mexico City negotiated content and themes on the *Moodle* MexCo forum with minimum intervention from their tutors. Other students decided to teach each other English or Spanish and design 'TEFL' online training units. This peer 'tuition' took place in the forum in the VLE, via email exchanges, via *Skype* exchanges and within the *Mahara* project designed by the 'budding' TEFL teachers (students following TEFL joint degrees). The TEFL tutor, who also works on the MexCo team (Elwyn Lloyd) monitored the teaching units that the students were preparing and briefed students on how to provide feedback. Students in both countries recorded their experience with metareflective logs.The collaborative task that students at CU had to create was assessed and accounted for fifty per cent of the module credits.

Issues relating to intercultural understanding (or misunderstanding) had been highlighted in the pilot phase of the project so it was decided to design a creative collaborative task because we believe that:

---

1 At Coventry University a BA Honours degree normally consists of three hundred and sixty credits, one hundred and twenty per year. Each student normally takes the equivalent of six twenty credit modules (roughly corresponding to subjects on their course) per year.

> in order to help students to cross threshold concepts it is necessary to devise student-centred activities that allow them to engage both in individual and collective reflection on the troublesome knowledge encountered. The overcoming of 'stumbling blocks' will be greatly helped by the opening up of a dialogue between students and tutors and amongst students themselves and by activities that foster this dialogue, as well as by encouraging students to engage in 'metareflection' on the difficulties encountered (Orsini-Jones 2008: 220).

The stated learning outcomes for the module *Introduction to Studying English and Languages at University* were that by the end of the module students should be able to:

1. Search for, review, cite and correctly reference literature using the Harvard style;
2. Interpret data from a variety of sources, evaluate and summarise them;
3. Plan and write an essay according to expected academic standards for university work;
4. Demonstrate the ability to work as part of an international team and design a collaborative digital online learning project.

The intended learning outcomes were assessed via a collaborative digital group project (outcomes 1, 2 and 4) and an essay (outcomes 1, 2 and 3).

The task brief given to the CU students for their digital learning object was the following:

> Some students from our partner University in Mexico (FESZ/UNAM) are planning to come to Coventry. Create a web-based Spanish/English (or English only) guide (a learning object) to Coventry/Coventry University for them using the e-portfolio *Mahara*. The guide must include both academic and non-academic information. You need to consult the students from FESZ/UNAM on its design via *Moodle* (international exchange) and provide examples of how their input (both content and design) informed your project. You may include jointly created blogs in the project and discuss what you learnt from the experience. You will also need to keep an individual reflective record of your intercultural exchange (in terms of personal feelings, new knowledge and actions that stemmed from it) using the *Mahara Project Journal* tool. (*Mahara* is an open source e-portfolio that can also be used to enable students to collaborate and create pages together in a "collection" that resembles the working of a *Wiki*. Further details available at: <https://mahara.org/>.)

Although task-based projects had been carried out before in the Department of English and Languages (Orsini-Jones 2010), these interactive learning objects created by the students at CU for their peers at FESZ gave a fascinating insight into what 'culture' is for our students. The learning objects also highlighted themes that had emerged in the asynchronous discussion threads in *Moodle*, that is to say that CU students' understanding and visualizing of culture appeared to differ considerably from the understanding and visualizing that were expressed by students in Mexico City (more in this below). On the other hand, students also discovered that they had many interests in common, for example a love for Gothic and horror literature.

The role-reversal model was embodied well by these projects, as staff gained an insight into the way their students understood culture through their intercultural digital learning objects. The task gave students the opportunity to work autonomously, display their creativity and enhance their digital literacies. In academic year 2013–2014, the students who designed the learning objects in 2012 became 'expert students', supporting their peers as mentors and in so doing drove forward the next action research cycle.

The project enabled the students at CU to:

- Find more opportunities to plan their discourse with the support of e-tools.
- Reflect on their production.
- Compare their production with that of their peers both at CU and in Mexico.
- Share their work with peers and lecturers as the URLs created in *Mahara* were posted in *Moodle* and viewed by all participants.
- Share a more democratic setting with their lecturers who became more like their peers in the online environment.
- Acquire and develop useful academic digital and transferable skills while at the same time making the most of and sharing their own digital knowledge (students included various Web 2.0 tools and software in their digital objects such as *YouTube*, *Flickr*, *Google+* and *Blogger)*.
- Negotiate meaning both with their Mexican peers and amongst themselves in their groups enabling them to become more interculturally aware.

It could be argued that constructing the learning objects gave the CU students the opportunity to "loosen their boundaries through accommodation" (Ackermann 1996: 33) and this enabled them to bring the troublesome portal of ICC "into view". Ackermann argues that "projection of self in context" (ibid.) is fundamental to learning via collaborative object construction. Drawing from work by Sayeki (1989), Ackermann states that "people often throw out pieces of themselves ... into the contexts that they try to understand. In doing so, they become able to feel situations 'from within'. They build an analog of the blind person's cane. This kind of "becoming other" is different from simply diving into a situation" (ibid.).

Data collected

Both qualitative and quantitative methods of data collection (mixed methods approach) were used for this action-research project (see Burns 2010 on action-research), with a stronger emphasis on qualitative data ('QUAL-quant' model, see Dörnyei 2007). BERA – British Educational Research Association – (2011) makes specific recommendations on voluntary informed consent in relation to action research:

> Researchers engaged in action research must consider the extent to which their own reflective research impinges on others, for example in the case of the dual role of teacher and researcher and the impact on students and colleagues. Dual roles may also introduce explicit tensions in areas such as confidentiality and must be addressed accordingly (p. 5).

The 'closeness' of lecturers with the students participating in action-research-driven curricular change could at times be perceived as threatening by students, particularly if, as in this case, the principal investigator is also the module leader. For this reason, when arranging the semi-structured interviews that informed this study, it was made clear to all students that the participation was voluntary and that their grades would not be affected in any way by participating (or not participating) in the interviews. Information sheets were also distributed in class both in Mexico and in the UK and consent forms were designed for each action-research intervention.

The planning stage involved the discussion amongst staff and students of how to create a dedicated Virtual Learning Environment (VLE) area where students and staff from the two institutions involved could exchange information through an asynchronous discussion forum (Moodle was used). This phase was deliberately left to grow 'organically' and informally and was not integrated into the curriculum, nor linked to assessment. The choice of using a proprietary system like Moodle as opposed to a Social Networking Site (SNS) like Facebook was deliberate, as previous research carried out at Coventry University (Orsini-Jones, Brick and Pibworth 2013) had shown that SNSs can create ethical challenges both in terms of research methodology and privacy. Students had however the freedom to communicate via other social media, which they did (e.g. Skype and Facebook). In the first phase (April-October 2012) there were 83 participants registered on the VLE, however the number of participants that actively posted on the discussion forums was 39, the majority of whom were from Mexico. In the second phase, (October 2012-December 2012), which was integrated into the curriculum, from the 140 UK students who were offered the opportunity to take part in the exchange 40 opted to do so, and in Mexico 33 from 60 did.
The data analysed consisted of:

Phase 1 (Pilot)

- Language data exported from the discussion forum.
- Questionnaire responses for questionnaires administered online via *SurveyMonkey*.
- The research journal kept by the 'expert student' via the e-portfolio *Mahara*.

The data that emerged were analysed by a Master's student for her dissertation (Gazeley 2013).

Phase 2
As above and:

- Semi-structured interviews carried out with self-selected students (13% of participants took part in the UK, only one from Mexico)
- Analysis of the collaborative learning object designed by students as part of their assessed tasks

- Analysis of the students' reflections on the experience (both in the UK and Mexico)

In both phases language data were exported into *Word* and *Excel* files and analysed qualitatively to identify salient themes. Data were also analyzed quantitatively via *Wmatrix* (University of Lancaster 2006) to classify the frequency of certain words and explore the themes further.

In Phase 2 semi-structured interviews were carried out, transcribed and coded using the research software *NVivo*. Demographic data such as gender, age, first language, institution of study, subject studied and language used in the discussion forum were coded numerically. The data were also submitted to frequency analysis via *Wmatrix*. This was done to analyse aspects such as frequency and key-word lists. In both phases a questionnaire was also used to gather the information regarding the experience of the participants who had actively posted in the discussion forums. The questionnaire was designed on the web with *SurveyMonkey*[2] (2013) and emailed out to the participants in English and Spanish versions. There were 28 questions for the participants to answer, including a variety of formats (ranking, open questions, yes-no) as recommended in the literature on questionnaire design (Dörnyei 2003). Nine statements were designed to ascertain the respondents' opinions or attitudes surrounding their experiences of taking part in the intercultural exchange. This set of question statements were based on the "A self-assessment of my intercultural experience" Council of Europe assessment criteria by Byram, Gribkova, and Starkey (2002: 26). In the pilot phase, 17 participants responded out of the 39 who were emailed (44% return rate). The responses for Phase 2 are being collated at the time of writing.

The discussion forum texts and the transcriptions of the interviews were also inputted into word clouds (*Wordle*[3] and *Tagxedo*[4]). Below are the word clouds for the discussion forums in the first two phases. They evidence how in the pilot phase the focus was on Mexico, while in the second phase the focus shifted to Coventry.

---

2 <www.surveymonkey.com>.

3 <http://www.wordle.net>.

4 <http://www.tagxedo.com>.

Figure 4: Word cloud (*Wordle*) of the text extracted from the discussion forum in Phase 1.

Figure 5: Word cloud (*Wordle*) of the text extracted from the discussion forum in phase 2.

The word cloud for the semi-structured interviews, that collated the transcription of five semi-structured individual interviews, illustrates how students are now reviewing their own initial perspectives. While nouns mainly emerged as the most frequent words from the discussion forum word clouds (Figures 4 and 5), it is quite powerful to see that words relating to reflection, cognition, collaboration and communication emerge in the most frequent words recorded in the interview transcripts, i.e. "think, "know", "people" and "talk":

Figure 6: Word cloud (*Tagxedo*) of the transcription of the semi-structured interviews with CU students carried out in May 2013.

It could be argued that the words were elicited by the interviewers, and their incidence is therefore not valid in research terms, but it was still interesting to see how consistently these terms emerged after the exchange.

In terms of the common interest themes identified, the table below illustrates the ones that emerged from the pilot phase and which were addressed subsequently in Phase 2 for task design purposes are the following (in order of popularity/incidence in the data from the pilot):

1. Culture/Traditions/Celebrations
2. Economy/Politics/Globalisation/History
3. Travel
4. Environment
5. Social Life/Society
6. Daily routine/Lifestyle

Some rather interesting comments were made by the Mexican students in the open questions in the online survey which gave insight into their "cultural frames" (Gumperz and Roberts 1991). The following citation for example was problematic for staff at Coventry University on a number of levels:

> "Es muy bueno poder hablar con personas de un país tan importante como Inglaterra y que esto me ayude a mi formación en el idioma es otro incentive para hacerlo." (our underlining)(It is excellent to be able to speak with people from a country as

> important as England and this helps me with learning the language and is another incentive to speak it)

First of all, although the UK is perceived to be very important, UK citizens can be embarrassed about the reasons why this is so. Secondly, one of the members of the project team is from Wales, and provided an explanation on the forum on the differences between England, Scotland and Wales, Great Britain and the UK and the intercultural implications of calling a Welsh person 'English'.

The students' reflective portfolios demonstrated however that the intercultural awareness of all students involved was growing and that many of them were overcoming their initial stereotypes – some British students thought Mexican students would be very different from them initially for example -. What emerged was the ability of the students from both countries to rethink the way they thought of "the other" and a healthy rethink of themselves and their own initial perspectives (British students' 'voices' from the semi-structured interviews held in May 2013):

> "I found an opportunity to see how other people, living in another country, think, live, the kind of interests they have. When you understand other cultures, you start understanding your own culture even more and also value it more"

> "Personally I have a lot of prejudices towards foreigners and I'm starting to change them, this is good for me".

> "It was interesting learning about their culture but also looking at ours from their perspective."

> "Before I didn't have a huge awareness of different countries and their cultures… you hear about them…but you don't have a first-hand experience of it from their own perceptions…"

As argued by Byam (1997: 34) and demonstrated through many other telecollaboration exchanges (e.g. see *Cultura* by Furstenberg and Levet 2010) it appeared that for many students the act of relativising self and valuing others was actually taking place and enabling students to suspend their judgment of others and to develop attitudes of curiosity and openness.

However, critical incidents did occur. It became apparent that, in line with the findings by Furstenberg and Levet (2010), the FESZ's students' understanding of what "culture" means can differ from that of the CU students, as clearly illustrated by the exchange in Figure 7.

Hi I'm
I am a Coventry University Student. Nice to meet you.
I am studying French and International Relations.

So we want to make a Blog about Coventry and the tourist attractions here.
Is there anything in particular that you want to know about coventry?

Speak to you soon!
Hasta luego mis amigos

Edit | Delete | Reply

HI!
by - Thursday, 15 November 2012, 08:01 PM

Hello !

What i shouldn't miss from coventry?, there are archeological zones?

I hope to hear from you soon,
Greetings from México!

Show parent | Edit | Split | Delete | Reply

Re: HI!
by - Sunday, 25 November 2012, 05:01 PM

Hi

No i mean things you like to do for fun like play football
Or maybe where you like to go for fun

Speak to you soon!

Ciao!

Figure 7: Critical incident in Phase 2 – *Moodle* Discussion Forum.

The above is an issue that stood out in the exchange, that is to say the frequency of enquiries from the UNAM students about the archeological sites, geography, history and gastronomy of Britain, while many of the CU students seemed more interested in finding out about nightlife, sports and television. On the surface it seemed as though the UNAM students had more of an appreciation for so-called 'high culture' while the CU students were interested in 'low culture'. This observation might call for a more thorough analysis of the socio-cultural background of the students in the two institutions involved.

It is probable that what the UNAM students were more comfortable writing about and asking about was influenced by their L2 vocabulary knowledge, and this could mean that sometimes they came across to the CU students as excessively formal due to a reliance on Latinate words. Again, this is an issue that could be explored with both sets of students to raise awareness of issues relating to pragmatic competence and of the impact that their forum postings may have on their interlocutors. Yus Ramos (2011) has defined 'cyberpragmatics' and 'cyberpragmatic competence' – skill in understanding others' intended meanings in computer-mediated communication, and in formulating messages that get across one's own intended meanings more effectively – is clearly of great importance in online exchanges and will be a very interesting area for future study. For example, on the use of English as a *lingua franca* on the Internet, Yus Ramos (2011) argues that there is a tendency for users to abandon their own 'intra-cultural' politeness strategies and adopt a default level of politeness based on hyper-polite Anglo-Saxon norms. If this is indeed the case then it would be a disappointing development for those concerned about 'linguistic imperialism'.

A study by Montero-Fleta *et al.* (2009) compared postings in online fora in English and Spanish and found that while in CMC generally there is more evidence for the informalisation of discourse in both languages, English postings tended to be shorter, more informal, and exhibited a greater tendency towards orality. This is just one small example to illustrate how people from different "languacultures" (Agar 1995) will be used to different norms when they engage in online intercultural interactions, just as they might in offline face-to-face interactions. It is notoriously easy for interlocutors from the *same* languacultural background to misunderstand each other's messages in context-poor computer-mediated communication. Many of us will have wondered at some point why an email is so blunt, so cold, so rude or so long, and even the opening greeting lines, terms of address, and closing salutations the sender uses can cause unintended irritation. We often find ourselves pondering or trying to guess what thoughts or intentions lie behind messages we read through CMC channels, and through telecollaboration exchanges such as MexCo we have opportunities to really explore what our partners' thoughts and intentions are and learn about how they prefer to go about managing interactions – online and offline. In this way we will be enhancing our Intercultural Communicative Competence and understanding of each other. It should be noted that critical incidents occurred in communication between staff too.

It also became apparent in the course of the project that the CU students who were more willing to actively engage in the forum and with the collaborative project were those who had travelled themselves to be students in the

UK. We could possibly infer from this that for them the threshold concept of ICC had been crossed at an earlier stage because they had decided to study abroad and had already encountered the 'cultural blocks' that can hinder effective cross-cultural understanding, while it could be argued that student B in the exchange above had not even approached the threshold of troublesome intercultural knowledge. However, that same student did try to 'repair' after no reply was posted by the Mexican interlocutor after his posting. In the attempt below, we can see how the CU student appears to have perhaps realised that his approach had not been ideal in terms of cyberpragmatics. In threshold concepts terms we could see his attempt at 'repair' below as 'liminality'. He is oscillating near the ICC threshold concept, but not mastered it yet, his language betrays 'mimicry', that is to say 'faking' understanding of the ICC discourse. The posting below sounds more perfunctory than meaningful:

> HI
>
> Whats the culture like in Mexico? What tourist attractions do you have there? Do you like football? I Love it!
>
> Some of my friends here at coventry university and I have created a website about coventry and the tourist attractions.
>
> We made it especially for you, the students in Mexico:
>
> <http://mahara.coventry.ac.uk/view/view.php?t=jaeJi1Ug6HG7XcmkEhZI>
>
> Let me know if you like it
>
> Enjoy!

The fact that student B above struggled to embrace the ICC discourse is not surprising, as relating to 'otherness' presents an identity challenge. The ontological shift inherent to becoming a global citizen requires a serious re-discussion of self and of one's beliefs. Davies (2006: 71) makes a clear link between the act of learning and that of identity formation. Land and Meyer (2010: 72) discuss how the transformation caused by encountering new troublesome knowledge can prove to be challenging for many students. He quotes Bayne (2008), who in turn draws on Kristeva (1991) and says that: "To teach […] is to engage with new ontologies which make *students strangers to themselves"* (p. 204) (italics in the original). The resistance of some of the CU students to truly engage with the international project could probably partly be explained by the fear of facing otherness.

On a more positive note, we received this feedback from one of the partners in Mexico (Bustos, email communication 12 December 2012):

> Promoting these types of intercultural communication helps students to overcome stereotypes about the cultures where the language they are studying is spoken as a native language and about the people who speak that language. In the case of my students they seem to be more confident using English, they seem to be more motivated because they want to be more fluent so that they can communicate with their new friends in England and in the same way they are happy that they have helped some Spanish students with problems in Spanish. Now learning English has a new meaning for my students because they can see a real purpose in learning this language. They have learned that people in general have similar needs.

Concluding remarks

Staff and students involved in this project agree that the intercultural communicative competence of all parties involved has been enhanced, as evidenced in the forum postings themselves and in the comments the students made during focus groups and interviews. One of the most valuable outcomes has been the realisation for UK-based students that their Mexican counterparts "are just like us" and "they like the same things" (semi-structured interviews May 2013), but they also do things "differently from us", for example the celebrations for the *Día de Muertos* ('Day of the Dead'). This reflects the social justice dimension that such an exchange can bring to the fore and the way in which it can help undergraduates to grow into global citizens able to recognize and value cultural difference. Several students in the exchange had not travelled outside of their home countries (two of the CU students did not actually possess a passport). Some students from CU admitted having held beliefs about the 'national character' of people from Mexico, influenced by stereotypes that are based on often inaccurate and usually negative inferences accumulated over the course of history, and perpetuated and reinforced through the media, films and television. Suffice to mention the way Latin American people are portrayed at the beginning of Spielberg's blockbuster *Indiana Jones and the Raiders of the Lost Ark* (Spielberg 1981) i.e. as treacherous gangsters (López-Peláez Casellas 2013). For ICC to be properly developed, intercultural communicators do need to "examine and challenge generalisations or stereotypes, and suggest or present other viewpoints" (Byram, Gribkova and Starkey 2002: 20).

On the theme of negative stereotyping, many authors have decried what they see as "cultural and linguistic imperialism" and the continuation of "an 'us/them relationship in which a Western Self imagines a culturally deficient foreign Other" (Holliday 2007: 364). While this particular statement seems to represent an extreme standpoint, the theme of 'othering' appears frequently in literature dealing with how language learners are inevitably exposed to the culture(s) associated with their target language. Guest (2002) criticises the "essentialising of the Other" (p. 154), that is reducing a culture (often taken as a synonym for 'nationality') to a few 'essences', or stereotypical totems. Unfortunately it may simply be human nature to do this, and while perhaps narrow-minded it is much more often a manifestation of naïveté than of anything more wilfully dismissive or disrespectful. It would be harsh to accuse the group of CU students who named a sub-forum in *Moodle* 'Fish, Chips and Tacos' of 'essentialising' (and at least they essentialised their own culture as well as the Mexicans'), when in fact what they were trying to do was reach out to the students in Mexico. However, the tutors should be in a position to exploit such occurrences and use them to give students guidance on the issues of essentialising and 'othering', and in this way help students improve their intercultural awareness.

On a positive note, there was great pride on both sides in sharing 'real' experiences. Some of the CU students were also taking part in another telecollaboration exchange at the same time as MexCo and commented that the whole experience of negotiating themes and presenting their digital artifacts to their Mexican peers without too much guidance from tutors had made them feel a stronger sense of ownership in the knowledge-sharing process on MexCo and had made them seriously commit to the exchange. It is undeniable that for these students the experience was transformative, confirming Vera López's (2012) belief that "the study of Modern Languages and Cultures should include the agentification of the learner to enable them to make sense of their experience in a foreign language and culture" (p. 8).

As for the staff involved, the project has triggered increased curiosity for intercultural and cross-cultural pedagogical epistemologies (and ontologies) and for the complexities inherent to "becoming a telecollaborative tutor". It has also highlighted the difference in e-learning approaches in different countries and the need to have a robust discussion on the integration and utilisation of technology in telecollaborative exchanges (see Dooly and Sadler 2013 on this point).

To conclude, this journey of intercultural discovery made all participants realise that more work needs to be done to properly embrace the social justice agenda and confront their own stereotypes and misconceptions. The road to a postmodern and ethical global model of digital ICC is definitely challenging, but also rewarding. In the light of the data collected and the critical incidents that have occurred, it would appear that ICC is a threshold concept for both staff and students.

References

Ackermann, E. (1996) Perspective-Taking and Object Construction. In: Kasai, Y. and Resnick, M. (eds.), *Constructionism in Practice: Designing, Thinking and Learning in a Digital World*, pp. 25–35, Mahwah, NJ: Lawrence Erlbaum.

Agar, M. (1995) *Language Shock: Understanding the culture of conversation*. New York: William Morrow.

Barnett, R. (1997) *Higher Education: A Critical Business*. Buckingham: Society for Research into Higher Education and Open University Press.

Bauman, Z. (2000) *Liquid Modernity*. Oxford: Blackwell Publishing.

Bayne, S. (2008) Uncanny spaces for higher education: Teaching and learning in virtual worlds. *ALT-J: Research in Learning Technology*, 16(3): 197–205.

British Educational Research Association (BERA) (2011) Ethical Guidelines for Educational Research. Retrieved 10 September 2011 from <http://www.bera.ac.uk/files/2011/08/BERA-Ethical-Guidelines-2011.pdf>.

Beetham, H. (2007) An approach to learning activity design. In: Beetham, H. and Sharpe, R. (eds.), *Rethinking Pedagogy for a Digital Age – Designing and delivering e-learning*, pp. 26–40, London: Routledge.

Burns, A. (2010) *Doing Action Research in English Language Teaching : A guide for practitioners*. New York: Routledge.

Bustos, F. (2012) Email Communication 12 December.

Byram, M., Gribkova, B., and Starkey, H. (2002) *Developing the Intercultural Dimension in Language Teaching: A Practical Introduction for Teachers*. Strasbourg: Council of Europe.

Byram, M. (1997) *Teaching and Assessing Intercultural Communicative Competence*. Clevedon: Multilingual Matters.

Carroll, L. (1871) *Through the Looking Glass, and what Alice Found there.* 1st edn. E-book, The Millennium Fulcrum Edition 1.7, Project Gutenberg produced by David Widger.

Castells, M. (2000) *The Rise of the Network Society*. 2nd edn. Malden, MA: Blackwell.

Cousin, G. (2009) *Researching Learning in Higher Education.* New York: Routledge.

Crystal, D. (1997) *English as a Global Language.* Cambridge: Cambridge University Press.

Davies, P. (2006) Threshold Concepts: how can we recognise them? In: Meyer, J.H.F. and Land, R. (eds.), *Overcoming barriers to student understanding: Threshold concepts and troublesome knowledge*, pp. 70–84, London: Routledge Falmer.

Department for International Development (2013) 'What is Global Education?' Retrieved 4 July 2013 from <http://webarchive.nationalarchives.gov.uk/20130128103201/http://www.dfid.gov.uk/get-involved/in-your-school/global-school-partnerships/about-gsp/what-is-a-global-dimension/>.

Developing digital literacies and intercultural communicative competence for the 'global graduate' – Computer Mediated Communication for Interaction in Language Learning (#CWIL) *Joint Conference Coventry University/University of Warwick. Coventry 26–27 June 2013.*

Dooly, M. and Sadler, R. (2013) Filling in the gaps: Linking theory and practice through telecollaboration in teacher education. *ReCALL*, Volume 25 (1): 4–29.

Dörnyei, Z. (2007) *Research Methods in Applied Linguistics*. Oxford: Oxford University Press.

Ellis, R. (2003) *Task-based language Teaching and Learning*. Oxford: Oxford University Press.

Fairclough, N. (1995) *Critical Discourse Analysis: the Critical Study of Language*. London: Longman.

Flanagan, M. (2011) Threshold Concepts: Undergraduate Teaching, Postgraduate Training and Professional Development – A short introduction and bibliography. *UCL Department of Electronic and Electrical Engineering*. Retrieved 10 January 2012 from <http://www.ee.ucl.ac.uk/~mflanaga/thresholds.html>.

Fraser, N. (1996) Social Justice in the Age of Identity Politics: Redistribution, Recognition, and Participation. *The Tanner Lectures in Human Values*. Delivered at Stanford University April 30– May 2. [PDF document]. Retrieved 12 July 2013 from: <http://tannerlectures.utah.edu/_documents/a-to-z/f/Fraser98.pdf>.

Freire, P. and Macedo, D. (1987) *Literacy: Reading the Word and the World*. South Hadly, MA: Bergin and Garvey.

Furstenberg, G. and Levet, S. (2010) Integrating Telecollaboration into the Language Classroom: Some insights. In: Guth, S. and Helm, F. (eds.), *Telecollaboration 2.0*, pp. 305–336, Bern: Peter Lang.

Gazeley, Z. (2013) *MexCo: Exploring Intercultural Computer-Mediated Communication (CMC) and Metacognitive Thresholds in Language Exchanges*. [Unpublished MA Dissertation]. Coventry: Coventry University.

Gumperz, J. J. and Roberts, C. (1991) Understanding in intercultural encounters. In: Blommaert, J. and Verschueren, J. (eds.), *The Pragmatics of Intercultural Communication*, pp. 51–90, Amsterdam: John Benjamins.

Gramsci, A. (1947) *Lettere dal Carcere*. Torino: Einaudi.

Guest, M. (2002) A critical 'checkbook' for culture teaching and learning. *ELT Journal Volume*, 56 (2): 154–161.

Guth, S. and Helm, F. (eds.) (2010) *Telecollaboration 2.0*. Bern: Peter Lang.

Guth, S. and Helm, F. (2011) Teaching Culture through CALL. In: Arnold, N. and Ducate, L. (eds.), *Present and future promises of CALL: From theory and research to new directions in language teaching*, pp. 211–256, San Marcos, TX: CALICO Publications.

Hafez, K. (2007) *The Myth of Media Globalisation*. Cambridge: Polity Press.

Helm, F. and Guth, S. (2010) The Multifarious Goals of Telecollaboration 2.0: theoretical and practical implications. In: Guth, S. and Helm, F. (eds.), *Telecollaboration 2.0*, pp. 69–106, Bern: Peter Lang.

Holliday, A. (2007) Response to 'ELT and "the spirit of the times"'. *ELT Journal*, 61 (4): 360–366.

Hokstad, L. M. and Dons, C. (2007) Digital literacy–towards a redescri tion of literacy for the digital learning environments. Paper presented at the Conference Interactive Computer Aided Learning [online] 26–28 September 2007. Retrieved 21 July, 2013 from <http://halshs.archivesouvertes.fr/docs/00/25/71/24/PDF/210_Final_Paper.pdf >.

Hutchinson, J., Atkinson, K. and Orpwood, J. (1998) *Breaking Down Barriers*. London: Stanley Thornes Ltd.

Klapper, R. (2006) *Understanding and Developing Good Practice: Teaching Languages in Higher Education*. London: CILT.

Kemmis, S., and McTaggart, R. (2005) Participatory Action Research: Communicative Action and the Public Sphere. In: Denzin, N.K. and Lincoln, Y.S. *(eds.)The Sage Handbook of Qualitative Research* (3rd edn) pp. 559–603, London: SAGE.

Kumaravadivelu, B. (2001) Towards a Postmethod Pedagogy. Retrieved 12 December 2012 from <http://bkumaravadivelu.com/articles%20in%20pdfs/2001%20Kumaravadivelu%20Postmethod%20Pedagogy.pdf>.

— (2008) *Cultural Globalization and Language Education*. New Haven, CT: Yale University Press.

— (2012) *Language Teacher Education for a Global Society*. Abingdon: Routledge.

Kristeva, J. (1991) *Strangers to ourselves*. New York: Columbia University Press.

Land, R., Meyer, J.H.F. and Smith, J. (eds.) (2008) *Threshold Concepts Within the Disciplines*. Rotterdam: Sense.

Land, R. (2011) Speed and the Unsettling of Knowledge in the Digital University. In: Land, R. and Bayne, S. (eds.), *Digital Difference: Perspective on Online Learning*, 61–70, Rotterdam: Sense Publishers.

Land, R. and Meyer, J.H.F. (2010) Threshold Concepts and Troublesome Knowledge (5): Dynamics of Assessment. In: Meyer, J.H.F., Land, R. and Baillie, C. (eds.), *Threshold Concepts and Transformational Learning*, pp. 61–79, Rotterdam: Sense.

López-Pelàez Casellas, M. (2013) *Teaching Film Studies: Discussion on Gender, Identity and Race in Higher Education*. 'Pedagogical Innovation in Languages and Literature (PILL) Seminar'. held 17 April at Coventry University.

Lloyd, E. (2012) Language Learners' 'Willingness to Communicate' through <Livemocha.com>. *ALSIC*, 15 (1) Retrieved 7 July 2013 <http://alsic.revues.org/2399>.

MacAskill, E. and Borger, J. (2013) New leaks show how US is bugging its European allies. *The Guardian*, 1st July: 1.

Mason, R. and Rennie, F. (2008) *E-Learning and Social Networking Handbook*. London: Routledge.

McLuhan, M. (1967) *The Medium is the Message: an Inventory of Effects*. London: Bantam.
McNiff, J. (1988) *Action Research: Principles and Practices*, London: Routledge.
Meyer, J. and Land, R. (2003) Threshold Concepts and Troublesome Knowledge: Linkages to ways of Thinking and practising within the Disciplines. *ETL Project Occasional Report 4.*
Meyer, J.H.F. and Land, R. (2005) Threshold Concepts and Troublesome Knowledge (2): Epistemological considerations and a conceptual framework for teaching and learning. *Higher Education*, 49: 373–388.
Meyer, J. and Land, R. (eds.) (2006) *Overcoming Barriers to Student Understanding: Threshold Concepts and Troublesome Knowledge.* Abingdon: Routledge.
MLA Ad Hoc Committee on Foreign Languages (2007) *Foreign Languages and Higher Education: New Structures for a Changed World.* The Modern Language Association of America.
Moon, J. (2004) *A Handbook of Reflective and Experiential Learning.* Abingdon: Routledge.
Montero-Fleta, B., Montesinos-López, A., Pérez-Sabater, C. and Turney, E. (2009) Computer-mediated communication and informalization of discourse: The influence of culture and subject matter. *Journal of Pragmatics*, 41: 770–779.
O'Dowd, R. (ed.) (2007) *Online intercultural exchange: An introduction for foreign language teachers.* Clevedon: Multilingual Matters.
O'Dowd, R. (2010) Online foreign language interaction: Moving from the periphery to the core of foreign language education? *Language Teaching Journal.* pp. 1–13. Retrieved 12 July 2012 from: <http://www3.unileon.es/personal/wwdfmrod/LTJ.pdf>.
— (2013) *The competences of the telecollaborative teacher.* [paper presented at Developing Digital Literacies and intercultural communicative competence for the 'global graduate' – Computer Mediated Communication for Interaction in Language Learning (#CWIL)'. Held 26–27 June at Coventry University and online via Languages@Warwick].
— and Ware, P (2009) Critical issues in telecollaborative task design. *Computer Assisted Language Learning,* 22 (2): 173–188.
O'Dowd, R. (2013a) The Competences of the Telecollaborative Teacher. *Language Learning Journal,* 41(3):1–14.

O'Dowd, R. (2013b) *The competences of the telecollaborative teacher.* 'Developing Digital Literacies and intercultural communicative competence for the 'global graduate' – Computer Mediated Communication for Interaction in Language Learning (#CWIL)'. Held 26–27 June at Coventry University and online via Languages@Warwick.

Orsini-Jones, M. (2008) Troublesome language knowledge: identifying threshold concepts in grammar learning. In R. Land, J. F. H. Meyer and J. Smith (eds.), *Threshold Concepts within the Disciplines*, pp. 213–226, Rotterdam: Sense.

Orsini-Jones, M. (2010) Task-Based Development of Languages Students' Critical Digital Multiliteracies and Cybergenre Awareness. In: Luzon, M. J., Ruiz, N. and Villanueva, L. (eds.), *Digital Genres, New Literacies and Autonomy in Language Learning*, pp. 197–224, Cambridge: Cambridge Scholars Publishing.

Orsini-Jones, M., Cribb, M., Jones, D., Graham., R. Lund, R. and Lee, F. (2010) Student-Driven and Threshold-Concept-Informed Curricular Change. iPED 2010. 5th International Inquiring Pedagogies Conference Proceedings, Coventry, September 15–16, pp. 77–83, Coventry: Coventry University.

Orsini-Jones, M., Brick, B. and Pibworth, L. (2013) Practising Language Interaction via Social Networking Sites: the "expert student's" perspective on personalized language learning in B. Zou (Ed.) *Computer-Assisted Foreign Language Teaching and Learning: Technological Advances*, pp. 40–53, Hershey, PA: IGI Global.

Orsini-Jones, M., Lloyd, E., Gazeley, Z., Bescond, G. and Lopez-Vera, B. (2013) Negotiating intercultural awareness and language varieties online: case study UK-Mexico [paper presented at the 5th COIL Conference: Can Globally Networked Learning Anchor Internationalization in the Curricular Mainstream? State University New York April 3–5 2013].

Orsini-Jones, M. (2013) *Towards a Role-Reversal Model of Threshold Concept Pedagogy*. Delivered at *Threshold Concepts: from personal practice to communities of practice – National Academy's Sixth Annual Conference and the Fourth Biennial Threshold Concepts Conference*, University of Dublin, Trinity College, June 28–29. Retrieved 26 August 2014 from <http://www.nairtl.ie/documents/EPub_2012 Proceedings.pdf#page=11>.

Perkins, D. (2006) Constructivism and troublesome knowledge. In: Meyer, J.H.E. and Land, R. (eds.), *Overcoming barriers to student understanding: Threshold concepts and troublesome knowledge*, pp. 33–47, London: RoutledgeFalmer.

Polanyi, M. (1958) *Personal Knowledge.* London: Routledge.

Ryan, P. M. (2002) Searching for the intercultural person. In: Alred, G., Byram, M. Fleming, M. (eds.), *Intercultural Experience and Education,* pp. 131–154, Clevedon; Buffalo; Toronto; Sydney: Multilingual Matters.

Sayeki, Y. (1989) *Anthropomorphic Epistemology*. Unpublished manuscript, Laboratory of Comparative Human Cognition. San Diego: University of California.

Sharifian, F. (ed.) (2009) *English as an International Language: Perspectives and Pedagogical Issues*. Clevedon: Multilingual Matters.

Schenker, T. (2013) *Intercultural Competence and Language Skills Through an ElectronicExchange.* 'CALICO 2013: Navigating the complexities of Language Learning in the Digital Age'. Held 21–25 May at the National Foreign Language Resource Center University of Hawaii, Manoa.

Siemens, G. (2004) 'Connectivism: A Learning Theory for the Digital Age'. elearning space: everything elearning. Retrieved 12 July 2013 from <http://www.elearn space.org/Articles/connectivism.htm>.

Shakespeare, T. (2007) *Disability Rights and Wrongs.* New York: Routledge.

Spielberg, S. (1981) *Indiana Jones and the Raiders of the Lost Ark.* LA: Paramount Pictures.

Thomas, G. and Loxley, A. (2007) *Deconstructing Special Education and Constructing Inclusion*. Berkshire: Open University Press.

Thorne, S. (2010) The intercultural turn and language leanring in the crucible of new media. In: Guth, S. and Helm, F. (eds) *Telecollaboration 2.0: Language and intercultural learning in the 21st century*, pp. 139–165, Bern: Peter Lang.

University of Lancaster (2006) *Wmatrix corpus analysis and comparison tool* [online] Retrieved July 2013 from <http://ucrel.lancs.ac.uk/wmatrix/ >.

Vera-López, H.B. (2012) *The Scholarship of Learning Modern Languages and Cultures: Integrating Education, Research and Human Development.* Unpublished PhD Dissertation. Nottingham: University of Nottingham.

Vygotsky, L.S. (1962) *Thought and Language*. Cambridge, MA: MIT Press.

Warschauer, M. (2002) <Languages.com:> the Internet and Linguistic Pluralism. In: Snyder I. (ed.), *Silicon Literacies: Communication, Innovation and Education in the Electronic Age*, pp. 62–64, London: Routledge.

—. and Grimes, D. (2007) Audience, authorship, and artifact: The emergent semiotics of Web 2.0. *Annual Review of Applied Linguistics,* 27: 1–23.

—. and Ware, P. (2008) Learning, change, and power: Competing discourses of technology and literacy. In: Coiro, J., Knobel, Lankshear, M. C. and Leu, D. J. (eds.), *Handbook of Research on New Literacies,* pp. 215–240, New York: Lawrence Erlbaum Associates.

Yus Ramos, F. (2011) *Cyberpragmatics: Internet-mediated communication in context.* Amsterdam / Philadelphia: John Benjamins.

Himanee Gupta-Carlson

Internationalizing U.S. Students with Hip-hop and Social Networks

> H.G. (instructor, United States): Being Canadian, being American. What's the difference?
>
> J.A. (instructor, Canada): There are some real differences between the two countries, but even so we have to also acknowledge regional differences within both countries.
>
> B. (student, U.S.): If you ask most Americans, the answer would probably be "not much" to the horror of our Canadian counterparts. But what is not understood by that perception is that many Americans feel Canadians are equal to them, and here in the U.S. we're supposed to be all about equality.
>
> G. (student, U.S.): "Equal but different" is a self-contradictory expression that can be applied to the American cultural comparison. One example is the American political parties posture for their positions on marriage equality, while our northern neighbors have found consensus.
>
> H.G. (instructor, U.S.): So if we were to turn these thoughts into a story, how would we start? Imagine yourselves to be rappers… You've got the mic.
>
> A. (student, Canada): Maybe the best place to start any story of a people, Canadian or American, is to talk about who they aspire to be. It seems only fair that a people be allowed to present their best selves, even if that self has not yet been fully realized.
>
> C. (student, Canada): Classified [a Canadian hip-hop artist] raps about how we need to be "real people" as opposed to comparing ourselves to others.

Canada and the United States share a long border, cordial governmental relations, and similar demographic, economic, and cultural profiles. These features might lead one to believe that Canadians and Americans, as neighbors, share many similarities. But, as the above excerpts from a co-created story project involving U.S. and Canadian college students illustrate, those similarities may only run skin deep. Furthermore, they may be obscured through a lack of education about each other and perhaps insufficient dialogue. Many American citizens know nothing about their across-the-border neighbors and might feel no reason to care. The situation differs for Canadians, most of who live within one hundred miles of the U.S. border and consume U.S. media, brand name foods, and fashions as other to the geopolitical power south. The United States, not by choice, is on Canadian minds often. As commentator William Petersen (2001) observes in an essay on Canadian-American relations, "A Canadian is likely to establish his identity by emphasizing the not-American part of himself and his nation. [...] to be Canadian means mainly to be different from, and possibly also alien to Americans" (p. 82).

Can this duality be ended? Can American college students gain consciousness and an interest in promoting international social justice through an experience of learning through conversations with Canadian students the politics, histories, and cultures of the two countries as well as the similarities and differences among themselves as individual residents of the United States and Canada? Can they teach and learn through dialogues via such vehicles of social media as the synchronous conversations over computer networks that constitute live chats, the social network spaces that establish a form of community through shared dialogue, or the self-developed web sites in which individuals or groups publish personal insights or thoughts like blogs?

A globally networked learning environment that I created at SUNY Empire State College in New York with Dr. Janni Aragon of the University of Victoria in British Columbia as part of a collaborative project through the State University of New York's Center for Online International Learning (COIL) offers one example of how educators in the humanities and social sciences who wish to expose American students to a more international environment might approach such questions. The vehicle we used to establish a common ground for students was the concept of hip-hop, not as a public commodity but as a practice of building community, sharing stories, and thinking about and mobilizing in politics. Students in my U.S.-based online Digital Storytelling class spent five weeks in 2012 discussing hip-hop practices and politics, and community building through storytelling and political activism via such electronic modes of exchange as e-mail, Facebook, Internet-based chat rooms, and a variety of other computerized social networking spaces with students in Aragon's Youth Politics seminar. While most of these conversations took place in groups of two to four students, both classrooms also came together over the five weeks in a shared Internet social networking space called StoryTimed to co-create the digital story project on the United States and Canada excerpted above. Through transcripts of their conversations, blog postings, discussion forum comments, and written essays, students pondered how these experiences had affected their impressions of community building and social justice, and logged ongoing reflections on the challenges and rewards that the cross-national communicative mode of learning was offering in discussion forums, personal blogs, essays submitted to their instructors, and in Aragon's case, in feedback via group face-to-face discussions.

This chapter describes how Aragon and I created our global classroom and how the learning experience impacted the American students. As my comments in the pages that follow will explicate further, I have chosen to focus primarily on the American students because the experience of being the American instructor who co-created and then actively participated in this international teaching collaboration sensitized me, too, to a sense of a perhaps unintentional but troubling blindness that can pervade classroom teaching philosophies and practices when one operates from the more dominant standpoint of a cross-national power relationship. Like my students, I entered the collaboration with an assumption that the U.S. and Canada, while different nations, held more in common culturally with each other than perhaps any other nation in the world might vis-à-vis the U.S. While the assumption holds weight, one realization that resulted from the downplaying of difference was a new awareness on my part of how such dismissals create conditions that enable articulations of diversity, internationalization, and cross-national relationships to be made primarily in the dominant power's terms.

Internationalizing with Social Networks

Recent scholarship on internationalization in higher education points to a growing interest among U.S. colleges and universities in using the Internet to create international learning experiences through partnerships with non-U.S. institutions (see Goastellec, 2010; Goodfellow and Hewling, 2005; Liaw and LeMaster, 2010; Matus and Talburt, 2009; Svensson and Whilborg, 2010; Ware and Kramsch, 2005). Traditionally, American institutions relied on such activities as study abroad programs to create such international experiences but, with the advent of new communicative technologies that allow students to converse via the synchronous computer networks of live chat, e-mail, Skype, and other Internet-enabled free or low-cost technologies, international collaborations are increasingly being established through distance-learning modes that allow students to access an international educational experience without having to take on the logistical burdens and financial costs of traveling abroad.

The rise of what often is called social media or Web 2.0 networked spaces is changing not only education but also the modes by which people create communities with each other (see Winograd and Hais, 2008). My own position is that of an American-born woman who has lived and worked her entire life in the United States. Growing up in this environment, I find even in a well-educated person such as myself less awareness of the world beyond the U.S. than I would like to admit, reflecting a commonly held belief that Americans are less accustomed to thinking of the world beyond their own nation's geo-political and cultural borders than are individuals who are citizens of other countries. At the same time, however, much has been said about the potential of the Internet and the burgeoning generation of Web 2.0 networks to overcome that American-centric disregard of the other. In my role as a faculty member and advisor at Empire State College, I have since 2010 mentored an arts student in Italy and a cultural studies student in India through the video-conferencing technology of Skype, and have taught U.S. history and digital storytelling online to students in China; the Czech Republic; and the United Arab Emirates. These students, like many others, complete their college degrees via the Internet, using the Internet based discussion forums, blogs, web linked resources, and e-mail to learn. Their experiences highlight how the social networks enabled by the Internet allow for an erasure of borders and illustrate their potential to educate.

A similar evolution in communication characterizes the hip-hop and youth political movements that Aragon and I used as topics and practices to put our students into contact with one another. Hip-hop gained popularity first in the Bronx and traveled by subway throughout New York, then by radio and TV across the U.S., and by cassette, video, and airplane all over the world. Students in both of our courses explored this evolution through Internet field trips to virtual hip-hop communities. These virtual communities are largely Internet sites that bring together artists, educators, and activists who are part of an underground (or non-commercial) scene. Individuals use the sites to exchange information on concerts, festivals, and workshops; to disseminate music and other art products created by themselves; circulate news on community events, post commentaries on the lives and experiences of artists; and mobilize hip-hop aficionados on issues ranging from the anti-corporate protest movements that began in New York City in September 2011 under the name of Occupy Wall Street and spread across the world to ongoing concerns internationally about marriage equality, sexually transmitted diseases, and human rights violations.

These sites show hip-hop to be a vital form of community building in both a highly localized physical sense as well as a globalized virtual sense. The local physical places and the global virtual spaces come together as a shared community through a joint understanding of what hip-hop is. For the artists, teachers, and community activists engaging with hip-hop, hip-hop is not merely the multi-million dollar entertainment industry whose best-selling commodity is commercial rap music. Hip-hop for the community based activists, artists, and teachers is a cultural movement defined by its five core elements: break-dancing, which is referred to as b-boying and b-girling in honor of hip-hop's roots in the Bronx as well as the beats created by breaks in music that many of the first hip-hop songs emerged from; emceeing, which is also known as rapping or performance with song and oratory; deejaying, which is the practice of creating musical beats through a manipulation of turntables and audio engineering technologies; graffiti, which is regarded as a vehicle for spreading messages of self-expressiveness via painting and words; and the fifth element, which is defined among hip-hop practitioners as self-awareness and personal growth through an acquisition of knowledge. I will explain the five elements in more detail below. For now, I would like to emphasize that these elements are practiced in both physical locales and virtual spaces. Break-dancing practices take place in churches, at gyms, in community centers, and when the weather is nice, outdoors at beaches and parks. Graffiti writing occurs legally in what hip-hop define as permission walls – public and private spaces where building owners or municipal governments have given artists permission to create murals and other self-expressive forms of art. Emcees and deejays perform, create music, and stage shows in coffeehouses, nightclubs, and restaurants as well as at churches, schools, community centers, and on outdoor stages in parks. But the products that hip-hop generates are shared, produced, and consumed largely through such social networks as

My Space, Facebook, Twitter, Tumbler, and similar no- or low-cost sites. In this capacity, hip-hop has evolved into an important educational outreach tool and community building practice among disenfranchised peoples who are finding ways to work together to overcome shared experiences of oppression (see Rose, 1994; Kitwana, 2003; Chang, 2005; Gupta-Carlson, 2010; Seidel, 2011; Foreman and Neal, 2011; Clay, 2012). The challenges that hip-hop artists, educators, community organizers, and other activists take on via social networking are aimed at furthering the education of a broadly defined global public in the hope of mobilizing through arts and activism an anti-racist, anti-classist agenda for a more socially just, egalitarian world. Hip-hop's power to reach a mass public has not been lost on progressive educators who increasingly incorporate a study of the movement into curricula and apply aspects of hip-hop's artistic elements of graffiti, break-dancing, emceeing, and deejaying to develop in students such important skills as self-articulation, public speaking, teamwork, personal confidence, physical well-being, and self-discipline. The use of hip-hop's artistic elements in curricula addresses the cultural movement's fifth element – the quest for self-awareness and personal growth through an acquisition of knowledge – and is the main rationale for the use of hip-hop in education.

Like hip-hop, which is a communal and democratizing activity committed to creating an inclusive environment that welcomes all, the learning communities that emerge through international educational tele-collaborations are said to be borderless in concept and equalizing in the expansion of learning opportunities to more students around the world. Drawing on the work done in foreign language education, for instance, Meei-Ling Liaw and Susan Bunn-Le Master (2010) note that through the development of abilities to speak a language other than one's own, students gain an opportunity to become not just multi-language speakers but also intercultural citizen activists. They use the term "intercultural" to highlight how the practice of learning a different language gives students "insight both into their own and the foreign culture and [establishes a need to be] aware of the meeting of cultures that often takes place in communication situations" (p. 21). They further argue that such students can learn to "engage with complex and multiple identities and to avoid the stereotyping that accompanies the perception of someone through a single identity" (p. 21). In doing so, the authors suggest, people of different nations are able to engage with the common medium of the Internet in modes of communication that "create compelling, surprising and sometimes even problematic conditions for learning" (p. 22) but nevertheless enable learners of different cultural, linguistic, and socio-economic backgrounds to "constitute themselves as a community, speak in a collective voice, converge on a linguistic style and concur on topics of conversation" (p. 22).

Amid the socially egalitarian promise that international education seems to offer, however, are sobering power differentials between educational institutions as well as the countries they represent. One of these challenges, identified by Claudia Matus and Susan Talburt (2009) in a discursive analysis of online learning spaces, is the dominance of Western institutions in the production of such collaborations and an accompanying acceptance of English as the main language of instruction. Arguing that universities not only respond to the logic of globalization that undergirds the growing interest in internationalization but also aid in its production, the authors notes that the hegemony of English as the primary medium for instruction and communication "privileges universities in industrialized nations as models for those in what are called nations 'in transition', developing nations, mid-sized economies, or small countries" (p. 517). This privileging replicates a form of colonial mimicry that forces the citizen of the less-industrialized, less dominant country to depend upon the morc-dominant partner. Richard W. Schwenz (2012) further notes the effects of such an imperializing practice in his commentary on internationalization in science education. Schwenz finds in science education courses in China, an aspiration to be like an American predominated to the point that instructors would conduct such courses primarily in English and speak wishfully of a day when science and science education in their country would be equal to that of the United States. In making this point, Schwenz wonders what values or potential contributions to science education might be missed. Is it fair, he asks, for one nation's norm to be teaching the world or would it be more egalitarian for all of us internationally to be contributing equally to that enterprise? This question of a lack of egalitarianism in awareness and knowledge confronted the American students who participated in the international collaboration that Aragon and I created when the Canadian students with whom they conversed pointed out, often quite bluntly, how their familiarity with the culture and politics of the United States outweighed that of the U.S. students' familiarity with domestic issues specific to Canada.

Paige D. Ware and Claire Kramsch (2005) further note that the primacy of discussion as a learning activity in online communicative learning environments is problematically Western and often regarded as a task that students perform for a teacher to evaluate rather than an opportunity to engage meaningfully in learning through conversation. In such scenarios, Robin Goodfellow and Anne Hewling (2005) suggest a dichotomy can arise, in which students accustomed to a Westernized norm of actively contributing to discussion are idealized as good students, independent and autonomous, while less active participants are categorized as lesser non-participants who are hurting not only themselves but the learning experience as a whole.

The privileging of English and Westernized norms of class participation in international telecollaborations occur in a context dictated by the norms of globalization, a force that has led to a commodification of education and arguably shapes the conditions under which U.S. and other Western institutions are investing in internationalization efforts. While the emancipatory potential that international collaborative efforts offer students should not be forgotten, maintaining awareness of the hegemonic dominance of the West in relation to the rest (Hall 1996) should not be undermined. The privileging of a Westernized – and increasingly an Americanized, masculine – voice can risk silencing non-Western styles of communication and stifle a potential diversity of voices that might exist in online environments.

Creating a Globally Networked Classroom

Aragon and I began the planning for our globally networked classroom a year in advance. Our initial development of the course began via conversations we held with each other by telephone and e-mail, and continued largely in that format as we began drafting an outline for the course in a learning management software program made available to us through the SUNY COIL Institute. Initially, we focused on two goals: Firstly, we wanted to develop curriculum out of an understanding of a pedagogy that was informed by our shared understandings of hip-hop as a community building and politically emancipatory practice; and secondly, we did not want to overwhelm or confuse our students by asking them to do work that would exceed the expectations of a standard undergraduate course. Elaborating on the second goal, we wanted our students' communications with each other to center not on memorizing or learning course material and more on gaining a better understanding of themselves and their cross-border counterparts as participants in a global society. We felt that this goal would best be accomplished if we drew on two key components of hip-hop's community building ideals: the establishment of knowledge building through conversations, and the use of a cipher, or unbroken circle, as a motif for stimulating conversations. The cipher is an organizing component used at hip-hop community events. As a circle of bodies, it is said to create a safe space where a performer can speak a truth, through bodily or verbal expression. The safe space, however, also is said to be a vulnerable space in that the performer is expected to speak with honesty and open-ness. If those values are violated, the individuals who constitute the cipher will use words or actions to chastise the performer.

Both Aragon and I are trained in feminist theory and political science, and believe in a principle of higher education as political praxis. We also both are professors at teaching institutions and have sought to create learning environments that allow for as much student input as possible. We value students' lived experiences and see them as contributing to the overall learning of all classroom participants. Our philosophies on teaching support an understanding of pedagogy as a philosophy and practice of theorizing and teaching from the perspective of knowledge gained at least partially through lived experience. Like the hip-hop educators whose models of teaching we sought to emulate, we generally strive in all of our classes to pay heed to history and to the importance of pushing against racist and colonialist practices of knowledge production that reify Western norms. Learning occurs in our classrooms not so much in lectures as in circles, where dialogue and a hearing of others is privileged over getting through a text.

Both Aragon and I also are aware of the unequal relationships of power that had informed the Western colonialist regimes of the nineteenth and early twentieth centuries as well as the continuation of such inequalities between more dominant and more subordinate global powers into the early twenty-first century and of the ways in which colleges and universities could participate in this perpetuation of these inequalities. We also are aware that while we are affiliated with educational institutions in industrialized countries where English is widely spoken, our institutional modes of education delivery and the demographics of our students differ. Aragon's position as senior instructor requires her to teach eight courses a year in all three terms of her university's academic calendar – fall, spring, and summer – and the learning occurs almost exclusively in face-to-face classrooms, where students are of "traditional" college age and reside on or near campus. My position involves teaching as well as the mentoring of students, the carrying out of independent studies, the development of curricula, and the supervising of adjuncts. I teach only four courses a year, and the learning occurs online through a course management system administered by my college. My students are mostly adult learners with a median age of thirty-seven. Most achieve their educational goals asynchronously, managing their course assignments and online participation around work, family, and other obligations, and rarely, if ever, coming into face-to-face or even voice-to-voice contact with instructors or their peers.

These differences – coupled with a three-hour time difference between our respective campuses – prompted Aragon and I to establish our respective courses as safe and separate spaces while using the communal practices of hip-hop as a modality for bringing our students together in social networks. We used a good neighbors metaphor to define our collaboration, reinforcing a point that each of our courses, like the United States and Canada, were places that housed students who would drop in to visit those in the other course, share ideas and readings, participate in activities together, and ultimately go back to their home course to report and be evaluated on what they had learned. Aragon adopted one of the required texts from my Digital Storytelling course for her students, and I replaced several general storytelling websites and YouTube video links that students usually would visit and discuss with online content on hip-hop communities and projects and political coalition building sites that her students also would view. We set up four shared assignments for our students to complete collectively, and I gave my students the option of completing a final digital story project with one or more of their University of Victoria peers. To facilitate the collective assignments, we created open-ended discussion prompts based on readings and materials related to hip-hop that both sets of students were assigned. To center the discussion further, we also asked students to read and discuss two essays on the political and cultural similarities and differences between the United States and Canada. I will discuss the various assignments and the level of student participation shortly. First, however, let us revisit the socially networked space StoryTimed with which I began this chapter and take a look at how the co-created digital story looked about mid-way through the collaboration:

> S. (student, U.S.): Canada lacks its own identity, not true. Canada suffers from younger sibling syndrome. Canada looks to the US to protect its borders, like a big brother. If the US is involved, it's easy to remain neutral, what will they have to lose. Big Brother will save us.
>
> A.(student, Canada): Canada's struggle to find a common identity often results in Canadians defining themselves as "not American." Our need to define our differences leads to us making sweeping stereotypes about Americans, something we accuse them of doing to us all the time. The ability to be identified as Canadian, or not an American … would likely be a compliment to many Canadians who take pride in being different.
>
> H.G. (U.S., instructor): Lovely thoughts and I appreciate the energy. How do we turn this all into a story?
>
> E. (Canada, student): I would have to disagree with the comment that "Canadians struggle to find their collective identity ... the US has a much stronger sense of self." Canada has not always been united. Divisions existed between Upper and Lower Canadas, among the provinces, and of course between our Indigenous Peoples and European settlers.
>
> F. (U.S. student): I would also have to disagree with the idea that Canadians struggle to find a collective identity. I live very close to the Canadian-American border and interact with many of these people every day. I may not be able to distinguish them by the clothes they wear, but what I do see is that many Canadians CAN distinguish the differences themselves. Whether there is a collective identity or not, from what I have seen in my own experiences, it does not seem to matter. They know who they are and what brings them together. I don't believe it's entirely a bad thing that they define themselves as "not American;" many cultures define themselves as "not this or that." While battles won and wars fought are certainly a way of creating nationalism and pride, it is not the only way.

Dialogue, dissension, and discomfort

Co-creating a digital story on the United States and Canada, which I entitled "Creating Community Through Story," was the first of the four assignments we asked our students to complete. Our goal was to apply the concept of a hip-hop cipher to an online environment. The cipher, writes hip-hop studies scholar Gwendolyn Pough (2004), is "a place where people gather to create knowledge and exchange information," (p. 41–2). Playing the role of a hip-hop emcee, I opened the story with the opening line – "Being Canadian, being American. What's the Difference?" I then asked the students in my course to practice a tactic of "call and response", which is used in both African American traditions as well as hip-hop culture, to generate a story together. Students were instructed to set up a free user account in StoryTimed and to send a friend request to me. Aragon, after friending me on StoryTimed in a similar fashion, posted a response to my opening line and asked her students also to set up accounts and to send friend requests to me. After I had "befriended" all of the participants, I set up a community group called Empire and Victoria and gave the students exclusive access to the story to co-create.

The rules of engagement within hip-hop are fairly open: Anyone who has something to express is welcome to speak as long as the cipher remains physically safe and verbally respectful toward all participants. The rules of our StoryTimed assignment were similarly open. Students were asked to contribute as they wished to the dialogue (although Aragon and I set a minimum contribution requirement of three), and, in the spirit of a campfire circle, to experiment with building a story by formulating their thoughts as responses to previous comments. We initially planned to let the activity run for two weeks but ended up continuing it through the full five weeks because student participation in the activity – while ultimately high – was initially delayed partly because of technological glitches with the networking space and partly because of slowness in the friend-request procedures.

In the hip-hop cipher, bodies a form a circle around a performer allowing the artist to dance, write, or speak a truth in a safe space that allows the performer to engage in self-expression. The circle of bodies protects the performer, and after the truth is expressed the performer returns to the circle, allowing the next participant to step into the middle. Although the space is safe, it is also vulnerable. For the performer is expected to drop all inhibitions and speak directly from the heart while the individuals forming the circle of bodies listen and respond with a goal of creating collectively through the individual expressions a body of knowledge, a story, song, or simply a dialogue that could lead to something new. That something new, however, may not be cohesive, comprehensible, or comfortable. And, as the collaboration evolved, it became increasingly clear that the story-in-process was not one of unity but of dissent, with the Canadian and American students rushing to defend their own individual positions:

> R. (U.S., student): I am mixed with many cultures Arab, Caribbean, Native, and African American but the thing that shocks people the most is that I'm a Muslim who covers or hides so to speak, her face and body. Or that I speak fluent Arabic, that I grew up listening to hip-hop, and R&B. That I talk with an American accent but the most shocking of all things is that I am "allowed" to continue my education although I am married with four children. This is my American dream, being me with no hindrance.
>
> K. (U.S., student): How many of us giggle when we hear someone use the term "A?" for instance and we immediately know that the person hails from Canada?
>
> N. (Canada, student): I must correct [student user name] and say that as a Canadian I am greatly offended by your misspelling of our national word. It's "Eh?" not "A?"
>
> L. (Canada, student): I find the American Dream to be an interesting phenomenon because it really pretends to be an equalizer for everyone. Perhaps this is to say it gives the illusion of a meritocracy. This is not to say that Canada in contrast is a better nation. We have our own struggles with equality. We do not have a Canadian dream that revolves around "making a name for yourself."
>
> M. (Canadian, student): I think that the idea of an "American Dream" is a bit ridiculous. By this I mean how should that dream be specific to America? Is it not true for most people is the world to want to be comfortable in life in terms of money, luxury, etc. I do not understand how this is something supposedly unique to America. I am sure that what constitutes success could be different from country to country, but success on its own is not a uniquely American desire.
>
> B. (U.S., student): The so-called American Dream is a misnomer. It (the dream) started as the hopes of oppressed peoples of Europe coming to the New World to free themselves from the oppression in their homelands. That dream has evolved over the centuries, but one thing has remained constant: people seem to believe the answer to their problems is in America.

The cipher is meant to stimulate dialogue and through dialogue establish a goal of forming a shared understanding between participants inhabiting a hip-hop space. Did such a shared understanding develop? Did a story take shape? I asked my Digital Storytelling students to consider these questions in an assignment to assess StoryTimed's effectiveness as a digital story tool. Their responses were generally negative. One student, for instance, observed, "I expected that we would collaborate and end up with one cohesive piece of writing that was not just a sounding board of 'what's wrong with whose country.' … instead, I found complaints against this country and a 'can you top this" mentality of Canada v. U.S.A." That feeling appeared to mirror the sentiments of Canadian students. Reporting on a conversation with Canadian A., American student G. noted that "the most revealing observation from [A.] shaped this comment: 'When we discussed it in class, many of my classmates were uncomfortable with the overt American exceptionalism coming from our US peers.' I can understand that statement since as a citizen experiencing the current presidential election rhetoric I sense the same national narcissism." Other U.S. students, however, felt the dialogue awakened in them a sense of how citizens of other countries perceive Americans and how much those perceptions are shaped not by the people but national governmental policies. U.S. student M. for instance, commented that "Some of the biggest points I realized in the StoryTimed project was how much stronger the moral values are in regards to religion in the United States." M. noted further that a statement from the story that "Canadians feel they need to assert themselves against American influence in an effort to preserve their own culture" helped him realize "how much our tradition, styles, music, movies, culture is being thrown across the border." This sentiment was echoed by American student R. who wrote in a final assessment that "we discovered through StoryTimed that the Canadian American relations need to be strengthened through vital communications and dialogue. The need for sensitivity and understanding was addressed through StoryTimed briefly, but more actions need to be taken."

The three remaining assignments in our global classroom matched U.S. and Canadian students with each other, and asked each group to converse three times – twice by e-mail and once in a synchronous mode such as telephone, Skype, or live chat. The Canadian students documented their conversations by posting a summary or a transcript to a blog in Aragon's course and the American students did so via a discussion forum in their course management system. Two of the conversations were focused on a discussion of the readings, web sites, videos, and other learning materials shared by students in both classes, while the third discussed the co-created story and the collaboration more generally. These conversations uncovered both personal similarities and differences between both the countries and the students themselves, with one American student, S. observing that

"I didn't realize the extent of strained relations between the U.S. and other countries until being involved in this collaborative assignment. I was met with extreme hesitation and vague answers. The student was timid and not willing to divulge any information. … 'I can't believe how friendly you are,' she said. When I read that opening line of an e-mail, I knew nothing was going to be easy." By contrast, American student K., an older adult learner, called her conversations with her Canadian counterparts "enlightening" but that "a twenty year old is almost always going to look at things and analyze things in a different way than a thirty-five year old. I've worked professionally and have been married and divorced while they're just getting ready to start their professional lives. We are going to have differing views."

Social Network Sites as Learning Spaces

In creating our global classroom, Aragon and I shared a fairly simple goal: We wanted to see what would happen if political science and humanities students in two separate countries started talking to each other. We wanted the students, in a sense, to take separate but complementary studies of storytelling and political activism and to work like hip-hop artists and activists to teach and learn from each other.

In addition to visiting StoryTimed, our students met with each other via e-mail, Facebook live chat, and another Internet-enabled synchronous social networking space called Today's Meet to discuss shared readings and their individual field trips to virtual hip-hop and youth political sites. For five weeks, they conversed in pairs or in groups of three to four on a variety of topics: The role that hip-hop activism played in the Occupy movements that had begun on Wall Street in New York City in 2011 and had spread to cities throughout the U.S. and the world, including many sites in Canada; the role that young Canadians adults were playing in a coalition organized as Stop the Cuts Toronto in mobilizing protests over federal policies; the upcoming November 2012 presidential election in the United States; differences between regional identities in their respective countries; differing views on such issues as health care and marriage equality; and their own personal hobbies and interests. The international collaborative project we designed within our two classes, in short, gave students in both of our classes an online venue in which to practice principles they were learning about community building and political organizing via hip-hop and digital storytelling.

As noted, I write from the perspective of a U.S. educator who teaches courses that are not explicitly international in their content. My courses, in other words, are aimed at studying histories, politics, and social concerns pertinent to the United States and not those of other nations. My project with Aragon of putting U.S. students in a storytelling course into contact with Canadian students in a political science course transgressed both disciplinary and national boundaries and in doing so called attention to a need to internationalize U.S.-centric curricula. From the collaboration, I gained a new desire to build in American students an awareness of and sensitivity toward people and concerns beyond the U.S. border, and feel that our global classroom offers one example of how the seeds for this awareness and sensitivity might be sown. The class met mostly in asynchronous socially networked spaces that existed outside both the physical walls of the University of Victoria campus and the virtual walls of the Empire State College learning management software system. Meeting in this virtual space, which increasingly has been referred to as the cloud, highlighted how learning through conversations with persons different from one's selves might expand the meaning of education beyond the fixed classroom itself. The conversations themselves connected American students (who often are perceived as ignorant to the world beyond U.S. borders) with students from other countries (who often are perceived as envious toward or resentful of their American counterparts) and illuminated how such practices might sensitize Americans to the problematic role of America's hegemonic position in the world and generate in them desires and ideas for creating cross-national alliances that might result in social change. At the same time, however, both the use of a virtual computer space outside the classroom and the conversations that ensued pointed to the challenges when one takes students out of the safe space of their own learning space to dialogue with others.

The goal of heightening student awareness and sensitivity toward matters of international justice was achieved in micro-level ways. American students reported throughout the five-week collaboration that they, often for the first time, had gained a new understanding of U.S.-Canadian relationships as well as a new sensitivity for the reasons why Canadians might be perceived as viewing Americans negatively. The students demonstrated this heightened awareness in discussion board postings, e-mail conversations with me, and in final self-assessments in which they were asked to reflect on the strengths and challenges of the collaborative experiment. These changes in perception suggest that Internet-enabled social networking tools such as the StoryTimed web site can bring students from separate classrooms, colleges, and countries together in lively, engaging, and meaningful conversations if the instructors integrate the learning activities effectively into overall course content. They also indicate that instructors can use asynchronous discussions via such modes as e-mail, Twitter, and StoryTimed to bring together students who reside and attend class in differing time zones, and that students respond well to such modalities. While synchronous discussions were difficult to organize on a classroom-wide scale because of differences not only in time zones but also in time demands on individual students, our encouragement to students to try and converse at least once synchronously did create opportunities for the American and Canadian students to engage more intimately with each other. When students could not directly converse in a synchronous mode, the use of synchronous tools such as Facebook chat and Today's Meet did allow for conversations to take place; those conversations were started, paused, and re-started over several hours. The ease of use of such electronic technologies made introducing non-American perspectives to course content a relatively simple process and generated positive responses from American students who often receive minimal exposure to histories, policies, and cultural practices of peoples outside America in courses that are not explicitly international.

Despite these successes, the power differentials in the relationships between the U.S. and Canada could not be erased. Those differentials need to be considered if such collaborations are to be sustained beyond one-time class-to-class experimentations. In the case of our tele-collaboration, a longer, sustained engagement would require me, firstly, as the U.S.-based educator to show how the reactions that the U.S. and Canadian students had toward one another can be contextualized within the relationships between the United States and Canada as well as in how the governmental and political leaders of the two nations position themselves vis-à-vis each other in a geo-political context as a way for U.S. students to gain understandings – and perhaps empathy – with the position of the so-called "other". On the non-U.S. end, the challenge would be for the educator to remind students of the class, race, gendered, and other inequities that divide peoples both within and across national boundaries. Even as non-U.S. students may be communicating with citizens of the global hegemon, the actual students who they encounter may very well occupy societal positions that are not equal to theirs.

Internationalization and social justice

None of the Empire State College Digital Storytelling students took up an invitation to create a story outside of the one in StoryTimed in collaboration with their University of Victoria Youth Politics partners. Two Empire State College students, however, chose to make the cross-border collaboration the theme of their final story projects. American student T. created a visual poster of words in a program called Wordle.org that captured the efforts from the student's standpoint to connect the challenges of working with another class to the relationship between digital storytelling and community building that the students had been studying in the course. In a commentary on the project, T. wrote, "Working with another class gave us the opportunity to experience, first hand, how to engage in digital storytelling. How to write and listen in a digital world in a more controlled setting … It wasn't a perfect collaboration. I read [in course discussion forums] how it worked for some of my classmates and not for others which tells you something." American student R. described the inspiration for a final project, also done in Wordle, as coming from a first conversation with the Canadian student that R. conversed with. R. wrote: "My initial conversation with UVIC student […] as well as the U-Vic collaboration deepened my understanding of the role of digital storytelling in our society. My creative process during the activity was trying to define such ideals of multiculturalism as a building up of the honesty of the individual while distinguishing that individuality is created in a range of social and cultural perspectives and viewpoints. In comparison to conventional ideas of citizenship, that stresses a separation of the individual and the state, multicultural citizenship deals with a trilogy of individuals, characteristics of people and the state.

These comments on the story projects suggest that if social justice is to be thought of as an educational practice, we must orient our teaching toward a better understanding of the world beyond our borders. These borders might be nationally, geographically, and/or culturally defined. In the case of this experimental collaboration, the borders became those that Aragon and I established in our thinking of the project as one of "good neighbors" who share ideas and visit each other but return ultimately to our own "homes". The "good neighbors" metaphor served a pragmatic function in one respect, in that it allowed students to be reassured that while they were sharing assignments, their "home" instructor would be the one assessing their work and evaluating their performance. At the same time, however, the metaphor had the side effect of not preparing students for the world of difference and discomfort that might exist outside their own safety zones. What none of us realized perhaps was that the classroom, unlike the "real world", was like the hip-hop cipher – a safe space, a community where people who had gotten to know and trust each other could feel free to speak their truths. Venturing into another classroom and being asked to interact with students who were younger (or older in the case of the University of Victoria students), at a more traditionally designed university (or a radically different one in terms of how Empire State College operates), and of another country (in the case of both sets of students) required all of us to leave the safe space. The challenges that the students encountered and the disgruntlement that was expressed were akin to the voices of new community organizers and social change agents who had not been adequately prepared for a world outside their borders. The value of this challenge was the fact that their own classroom existed and they could return to it to share their reservations and misgivings. What might have stretched the learning further would have been discussion of how to bring that world outside the classroom closer and more safely into the learning community that for each group of students had become their own.

References

Chang, J. (2005) *Can't Stop, Won't Stop: A History of the Hip-Hop Generation*. New York: Picador Books.

Clay, A. (2012) *The Hip-Hop Generation Fights Back: Youth Activism and Post Civil Rights Politics.* New York: NYU Press.

Foreman, M. and Neal M. A. (eds). (2011) *That's the Joint: The Hip-Hop Studies Reader*. New York: Routledge.

Goastellec, G. (2010) Small world: Access to higher education between methodological nationalism and international organizations. *Globalisation, Societies and Education,* 8(2): 283–293.

Goodfellow, R. and Hewling, A. (2005) Reconceptualising Culture in VirtualLearning Environments: from an "essentialist" to a "negotiated" perspective. *ELearning* 2(4): 355–67.

Gupta-Carlson, H. (2010) Planet B-Girl: Community Building and Feminism in Hip-Hop. *New Political Science* 32(4): 515–29.

Hall, S. (1996) The West and the Rest: Discourse and Power. In: Hall, S, Held, D., Hubert, D. and Thompson, K. (eds.), *Modernity: An Introduction to Modern Societies*, pp. 184–227. London: Blackwell Publishers.

Kitwana, B. (2003) *The Hip-Hop Generation: Young Blacks and the Crisis in African American Culture*.New York: Basic Civitas Books.

Liaw, M.-L. and Bunn Le Master, S. (2010) Understanding telecollaboration through an analysis of intercultural discourse. In *Computer Assisted Language Learning* 23(1): 21–40.

Matus, C. and Talburt, S. (2009) Spatial imaginaries: Universities, internationalization, and feminist geographies. In *Discourse Studies in the Cultural Politics of Education* 30(4): 515–27.

Petersen, W. (2001) Canadian-American Relations. In *Society*, pp. 78–82.

Pough, G. (2004) *Check It While I Wreck It: Black Womanhood, Hip-Hop Culture, and the Public Sphere.* Boston: Northeastern University Press.

Rose, T. (1994) *Black Noise: Rap Music and Black Culture in Contemporary America.* Middletown, CT: Wesleyan University Press.

Runnell, M. (2007) *The Hip-hop Education Guidebook.* New York: Hip-Hop Association.

Schwenz, R. W. (2012) Who is Catching Up With Whom? Internationalization of Science Education. In *Journal of College Science Teaching*, 41(3): 10–11.

Seidel, S. (2011) *Hip-Hop Genius: Remixing High School Education.* New York: R&L Education.

Svensson, L. and Whilborg M. (2010) Internationalising the content of higher education: the need for a curriculum perspective. In *Higher Education* 60: 595–613.

Ware, P. D. and Kramsch, C. (2005) Toward an Intercultural Stance: Teaching German and English through Telecollaboration. *The Modern Language Journal* 89: 190–205.

Winograd, Morley and Hais D. M. (2008) *Millennial Makeover: My Space, YouTube, and the Future of American Politics*. Rutgers, NJ: Rutgers University Press.

Nataly Tcherepashenets

Conclusion

Telecollaboration has the unique potential to enrich life and studies in a globalized world. The authors in this volume have explored the various ways in which telecollaboration can expand learning opportunities and contribute to the pursuit of equal educational outcomes among students around the globe. Both goals are closely related to the empowerment of learners and the social justice cause, inherent to the mission of higher education.

The contributors convincingly demonstrate that, despite challenges, telecollaboration can improve access to quality academic offerings, create new study opportunities, and produce learning environments conducive to the acquisition of intercultural communicative competency and creativity, arguably some of the most important skills for success in the 21st century.

The telecollaborative experiences discussed in this book shed light on desirable characteristics of academic offerings that are suitable for the online international classroom. These include challenging, thought- provoking content that is discussed in a context relevant for the development of in-depth understanding of both local and global realities (see, for example, Chapters One and Six). These characteristics feature awareness and sensitivity to linguistic and cultural differences, including barriers that can be used productively in negotiating meaning while celebrating the pluralism of perspectives that is intrinsic to the international classroom as it is to life in the global world (see, for example, Chapters Two and Eight). In addition, they supply a user-friendly course design that encourages active participation in a variety of asynchronous and synchronous modes in the spirit of solidarity, tolerance and open-mindedness (Chapter Three and Chapter Five). Guidance comes from instructors who are experts in their respective fields, as well as life-long learners and creative thinkers with cosmopolitan outlooks and sensibilities, and are eager to serve as role models for their students.

The authors also demonstrate at least three ways in which telecollaboration can advance social justice by removing barriers to educational access. First, in a spirit of solidarity and shared responsibility, telecollaboration can meet a global shortage of specialists in particular fields by ensuring the presence of local specialists in response to local need (for example, Chapter Three). Second, telecollaboration demonstrates how to bridge gaps in learners' preparation for competitive job markets, where increasingly there is emphasis on intercultural communicative competence in hiring and training. Online intercultural exchanges provide unique opportunities to develop this competence without the requirement of travel (see, for example, Chapters One and Five). Finally, as a vivid illustration of Freire 's (1998) view of education as a political act whose vocation is to address itself to "dreams, ideals, utopias, objectives" (100), telecollaboration can help to raise awareness about each individual's responsibility in the world and for the world's well-being, challenging global racism and exceptionalism through motivating learners to stay informed, demonstrating the relevance of global themes for local contexts, and enhancing civil engagement and solidarity in learners (see, for example, Chapters One and Nine).

The advancement of the social justice cause in education has been historically related to the empowerment of individuals. Telecollaboration can enrich understanding of the challenges and rewards of globalization as well as enhance the development of cosmopolitan outlooks and sensibilities, indispensable in our view for the success of globalization and for the life of democracy in the contemporary world, with its historically cherished values of tolerance, solidarity and social justice. These sensibilities can play a key role in the formation of the relational capital (see, for example, Chapters Four and Six), which opens numerous opportunities to experience and contributes to globalization while fostering students' and instructors' professional and personal growth. The accumulation of this capital also displays teaching and learning in telecollaborative setting as acts of solidarity which celebrate collaboratively created knowledge and shared resources, including open education tools (see, for example, Chapter Four). Allowing its participants to benefit from the liberating power of technology, telecollaboration, which balances local and global through personalized, individualized approach to learning that is reinforced through the presence of local instructors with global orientation, prepares learners for productive participation in the international dialogue, at the same time respecting differences and building on common goals.

By bringing authentic, often underrepresented voices from absence to presence in this dialogue, telecollaboration creates new learning opportunities, becoming an innovative stage for transformation for both students and instructors. Mezirow (2000) views the fostering of critical reflection on assumptions which learners and others have as a key element in perspective transformation, and he observes that transformative learning may lead to a fuller realization of human capabilities(see also this volume Chapters Six, Seven and Nine). Moreover, transformation achieved through intercultural learning can play a determinant role for the development of representational thinking. Indeed, as Hannah Arendt (2000) perceptively observes, the more voices are present in one's mind, the stronger representational thinker the individual is. We contend that the sharpening of representational thinking goes hand in hand with the development of intercultural communicative competence and creativity in telecollaborative setting, and that this is one of the most desirable educational outcomes of intercultural exchange. We view dialogicity, which Wegerif (2007) defines as an ability to respond to the other in a way that takes into consideration how one assumes others are going to respond, and the mental simulations of possible futures, or prospectus (Seligman et.al 2013), which Seligman considers to be an essential feature of human mind, as interrelated abilities which are at the core of the development of both intercultural communicative competence and creativity that inform representational thinking in telecollaboration. While such scholars as Byram (1997), Corbett (2010), and O'Dowd (2012) have already started productive study of the benefits which telecollaboration can bring to the development of intercultural communicative competence, its potential impact on the development of creativity, is a subject for future research. Both skills are in high demand at our conceptual age, which Pink (2008) views as one where the world economy and society rely on empathy, meaning-making, and big-picture capabilities, and, we may add, representational thinking.

We view these capabilities as requiring continuing renewal, which can be achieved through life-long learning. Online intercultural exchange unravels the importance of life-long learning through bringing awareness about inter-cultural connectedness (see, for example, Chapter Seven), motivating learners to stay informed about the world, and underscoring the relevance of the global themes for local contexts. With its potential to infuse civil engagement and activism across disciplines, telecollaboration defies the view of learning as an isolated practice and opens numerous unique opportunities for professional and personal development for instructors and students alike (see Chapters Two and Four). It also can serve as a vivid illustration of the positive impact of diversity on learning (Vygotsky (1997), Dooly (2008), see also Chapter Seven), in addition to shedding light on benefits of international team-teaching, which requires flexibility, respect for one's partner's teaching style, and open-mindedness, and can be enhanced through the highly desirable, in teacher training, completed in advance (see, for example, Chapters Two and Seven).

Thus, telecollaboration has a strong potential to improve and motivate learning, to unravel richness and complexities' of one's own culture and its relationships with others, and to foster intercultural understanding in the increasingly intertwined world. There are at least four conditions necessary for the realization of this potential:

1. Creation of a student-centered learning environment which encourages communication on equal terms and celebrates a variety of learning styles and backgrounds as opportunities for fostering intercultural understanding, acceptance of the plurality of perspectives, open-mindedness, and the questioning a single notion of truth (see, for example, Chapter Five).
2. Guidance by experts who are eager to embrace the adventurous side of education and who see themselves as life-long learners and agents of globalization, exemplifying flexibility and willingness to be challenged and changed by collaborative experience (see Chapters Four and Six).
3. Achievement of the balance between learning as a personal act and as a collaborative undertaking (see Chapters Eight and Nine).
4. Institutional support that includes the recognition of the importance of internationalization of education as a part of the mission of higher education, the willingness to share resources, and active engagement in the creation of open education tools (see, for example, Chapter Four).

With these conditions met, we suggest that the building of the culture of trust is indispensable for the ultimate success of telecollaboration. This culture requires taking risks to open oneself to others, an ability to challenge one's own way of thinking, and a desire to find a language in common through negotiation of meaning across cultural and linguistic differences. As in any setting, the building of this culture can be a challenging task. It is imperative, nevertheless, for the instructors who see themselves as agents of globalization and life-long learners to face this challenge with confidence and enthusiasm, as trust is the key to both the successful intercultural communication across virtual and real borders and to the life of the democracy in a globalized world.

It is hard to map the world as the consensual place; likewise, harmony does not always reign in the telecollaborative classroom (see, for example, Chapters Six and Nine). However, as musical theory reminds us, dissonance informs musical thinking and is integral to the transition to harmony. The buildup and release of tension are partially responsible for what listeners perceive as beauty, emotion, and expressiveness in music. The same, we suggest, is true for teaching, where conflicts inform pedagogical thinking (see, for example, Chapters Five, Eight, and Nine). Teaching can enhance and sustain trust through engaging learners in discussion of complex and controversial topics, encouraging self-confidence and open communication, as well as willingness to be challenged and changed through the personal example. With its potential to undermine any attempt to impose any dominant form of categorizing people, telecollaboration opens doors for the realization of Freire's (1998) vision of history as time filled with possibility, where the future is problematic, and not pre-determined, where each individual is responsible for accessing opportunities through network connections that increase his or her potential for social mobility and for moving out of a disadvantaged position. In conclusion, while enhanced by the liberating power of technology and competent instructors, telecollaboration advances understanding of the mission of education as a generator of confidence and agency in the world, which is both universal and fragmented, and whose well-being is dependent on each individual as potentially equal, eager and competent contributor in intercultural dialogue.

References

Arendt, H. (2000) Truth and politics. In: Baehr P (ed.), *The portable Hannah Arendt*, pp. 545–575, New York: Penguin.

Freire, P. (1998) *Pedagogy of Freedom. Ethics, Democracy and Civil Courage*. Trans. P. Clarke. New York: Rowman and Littlefield.

Corbett, J. (2010) *Intercultural language activities. Cambridge Hand books for Language Teachers.* Cambridge: Cambridge University Press.

Byram, M. (1997). *Teaching and Assessing Intercultural Communicative Competence*. Clevendon: Multilingual Matters.

Dooly, M. (2008*) Telecollaborative Language Learning*. Bern: Peter Lang Publishing Group.

Mezirow, J. and Associates. (eds). (2000). *Learning as Transformation*. San Francisco, CA: Jossey-Bass.

O'Dowd, R. (2012) Intercultural communicative competence through telecollaboration. In: Jackson, J. (ed.) *The Routledge Handbook of Language and Intercultural* Communication, pp. 340–356, Abingdon: Routledge.

Pink, Daniel. (2006) *A Whole New Mind: Why Right –Brainers will Rule the Future.* New York: Riverhead Books.

Seligman, M.E. P., Railton, P., Baumeister R.F., and Sripada C. (2013) Navigating into the future or driven by the past. *Perspectives on Psychological Science*, 8(2): 119–141.

Wegerif, R. (2007) *Dialogic Education and Technology*. New York: Springer.

Index

Access 19, 21, 22, 35, 38, 39, 40, 42, 43, 44, 45, 46, 74, 85, 93, 112, 119, 125, 133, 134, 137, 204, 205, 246, 254, 257, 258, 261
Action research 143, 151, 207, 211, 214, 215, 226, 229, 230
@genda 2.0 129, 132, 141, 142, 147–148
Aldrich, C. 39, 51
Al-Khatib, H. 35–52
Alred, G. 26, 29, 232
Amirault, R.J. and Vasser, Y.L. 29, 148, 149
Anderson, W. 177–201
Arab Open University 38, 41–42
Archer, L. 21, 29
Arendt, H. 259, 261
Bakhtin, M. 19, 29, 56, 58, 60, 62, 68, 70, 77, 79, 80, 81, 82
Bauman, Z. 27, 29, 61, 80, 81, 202, 226
Bax, S. 37, 39, 52
Belz, J.A. 179, 199
Bescond, G. 201–234
Bhagwati, N. 38, 52
Block, L. 35, 38, 40, 41, 52
Bourne, J. 38, 41, 52
Byram, M. 26, 29, 177, 178, 182, 199, 205, 206, 207, 217, 224, 226, 227, 232, 259, 262
Castells, M. 20, 22, 29, 203, 227
Chapelle, C. 35, 52
Charry Roje, R. 151–176
Connectivism 139, 140, 232
Competence 24, 25, 27, 28, 30, 50, 57, 62, 73, 74–75, 82, 127, 132, 135, 145, 147, 148, 149, 155, 156, 165, 178, 82, 199, 202, 203, 204, 205, 206, 207, 210, 212, 222, 224, 227, 230, 231, 232, 258, 259; pedagogical competence 72; communicative competence 181, 198, 259, 262; digital competence 202; cyber-pragmatic competence 25, 202, 209, 222, 223; intercultural communicative competence 30, 82, 178, 181, 182, 198, 200, 205, 222, 227, 258, 262
Corbett, J. 13, 177, 178, 181, 199, 259, 262
Creativity 31, 145, 214, 257, 259
Culture 19, 55, 56, 57, 58, 60, 61, 68, 75, 76, 78, 97, 98, 103, 116, 118, 119; culture and the concept of face 62–64
(inter/multi) cultural awareness 10, 25, 48, 72, 76, 88, 89, 93, 94, 95, 96, 97, 99, 100, 129, 154, 158, 178, 201, 206, 210, 212, 220, 225, 231
Daloz, L. 21, 29
Dewey, J. 20
developing countries 39, 40, 87
Dialogue 20, 24, 26, 56, 58, 66, 67, 71, 73, 75, 76, 78, 103, 113, 123, 165, 195, 205, 213, 236, 245, 246, 247, 248, 250, 258, 259; intercultural dialogue 35, 124, 147, 179
Diderot, D. 19
Digital Storytelling 116, 236–238, 247, 250, 252
Dooly, M. 24, 25, 30, 36, 52, 67, 81, 227, 260, 262
digital divide 22, 38, 39, 52, 102, 103, 173
education international 38, 39, 41, 87, 151, 238, 240, 241; without borders 27, openness in 29, 79, 112, special 28, 85–104, 232, distance 85, 102, 104, 105; empowerment through 22, 227, 258; educational outcomes 19, 257, 259; education as a political act 20, 236, 249, 258; open education 22, 24, 38, 41, 43, 44, 50, 51, 52, 104, 111, 112, 115, 124, 125, 135, 139, 152, 157, 206, 213, 217, 258

e-learning 21, 29, 35, 39, 41, 43, 44, 50, 51, 211, 225, 226, 238
exceptionalism 24, 248, 258
Facebook 72, 120, 121, 122, 123, 181, 211, 216, 236, 240, 249, 251
Freire, P. 20, 22, 30, 203, 228, 258, 271
Friedman, J. 20, 28
Gazeley, Z. 201–234
Guth S. and Helm F. 22, 25, 30, 67, 70, 81, 202, 204, 206, 207, 228, 232
Gupta-Carlson, H. 235–257
Goodfellow, R. 30, 36, 53, 237, 242, 254
Heidegger, M. "The Question Concerning Technology" 23, 24, 25, 30; "Origin of the Work of Art" 23
Husserl, E. 132, 149
Identity (construction of) 17, 21, 50, 68, 177, 180, 184, 185, 186, 188, 189, 194, 195m 196m 203, 205, 206, 208, 209, 223, 236, 240, 245
Inclusion 9, 13, 28, 38, 40, 44, 96, 100
Innovation 25, 29, 35, 36, 39, 40, 51
Internationalization (in education) 21, 25, 26, 28, 38, 43, 86, 102, 151, 237, 241, 242, 252, 260
Interlanguage 143
Jackson, S. 27, 30, 31, 82
Kant, I. 20, 30
King, K. 85–107
Kumaravadivelu, B. 25, 30, 202, 204, 206, 229
Kramsch, C. 126, 127, 237, 242
Knight, J. and Altbach, P.G. 21, 30, 35, 38, 39, 53
Kolb, A. 67, 68, 82
Kulavuz-Onal, D. 109–128
Lamy, M. 30, 36, 53
Learning (collective) 37, 42; (collaborative) 20, 38, 216; (experiential) 67, 82, 230;
language learning 26, 57, 110, 118, 122, 123, 126, 160, 177, 198, 204, 207, 209, 210, 211; life-long learning 24, 259
Lloyd, E. 201–234
López, B. 201–234
Martins, D. 151–176
Mcpherson, R. 85–107
Mezirow, J. 29, 259, 262
Mignolo, W.D. 24, 30
Milson, A. 85–107
multicultural awareness 88, 89, 94, 95, 96, 97, 99, 100
Networking 72, 92, 129, 140, 216, 229, 231, 236, 237, 240, 246, 249
Netnography 110, 114, 127, 128
O'Dowd, R. 21, 24, 25, 26, 30, 57, 58, 60, 68, 82, 83, 178, 198, 199, 200, 201, 203, 204, 206, 210, 230, 231, 259, 262
Orsini-Jones, M. 201–234
Papacharissi, Z. 22, 31
Participatory thinking 70, 77
Pelletier, C. 23, 31
Pedagogy (critical) 20
Representational thinking 259
Schmidt, R. (and the Noticing hypothesis) 70, 83
social justice 9, 10, 13, 17, 19, 20, 21, 24, 26, 28, 29, 37, 38, 43, 44, 55, 56, 60, 66, 79, 80, 87, 104, 154, 156, 157, 160, 161, 170, 185, 202, 204, 224, 226, 228, 236, 252, 253, 257, 258
solidarity 20, 21, 27, 28, 29, 48, 80, 82, 257, 258
special education 28, 85, 87, 88, 102, 104
teacher education 86, 202
Seligman, M. 259, 262
Siemens, G. 140, 232
Social networks 27, 29, 30, 82, 128, 129, 235, 236, 237, 238, 239, 244
Sousa Santos, B. 21, 31
Stetsenko, A. 20, 31
Tcherepashenets, N. 24, 31, 95, 96, 97, 99
Turula, A. 55–83
UNESCO 37, 41, 54, 87
Vásquez, C. 109–128
Vinet, D. 129–150
Vygotsky, L. 31, 211, 233, 260
World citizenship 19, 21, 28, 130
Webheads in Action (WiA) Online Community of Practice 110, 111, 112, 113, 114, 115, 116, 125, 128

TELECOLLABORATION IN EDUCATION

Edited by
Melinda Dooly & Robert O'Dowd

The series' focus is on the pedagogical processes and outcomes of engaging learners in different geographical locations in virtual contact and collaboration together. This contact can take place through the application of online communication tools such as e-mail, synchronous chat and threaded discussion as well as the tools of Web 2.0: like wikis, blogs, and online publishing. The series is also particularly interested in innovative teaching practices involving telecollaboration that integrate the use of newly emerging forms of Internet tools such as social networking or 3D virtual worlds.

A major aim is to reflect the diversity of research and practice in this area of knowledge, providing a space for transversal dialogue among teachers and teacher trainers, administrators, researchers, and educators working in different subject areas as well as various areas of education.

Telecollaboration in Education deals with the application of such activity in different subject areas (e.g. Foreign Languages, History, Science) and in different educational contexts, including but not limited to primary, secondary, university and adult education. Publications within the series include scholarly monographs and collected papers editions as well as cutting-edge projects that exemplify good practice in the application of distanced collaborative efforts. Training manuals for educators in the organisation and application of telecollaboration are also a possible type of publication within the series. Language of publication is English.

Vol. 1 Sarah Guth & Francesca Helm (eds)
Telecollaboration 2.0. Language, Literacies and Intercultural Learning in the 21st Century.
2010. ISBN 978-3-03943-0440-5

Vol. 2 Randall Sadler
Virtual Worlds for Language Learning. From Theory to Practice.
2012. ISBN 978-3-0343-0573-0

Vol. 3 Melinda Dooly & Robert O'Dowd (eds)
Researching Online Foreign Language Interaction and Exchange.
2012. ISBN 978-3-0343-1155-7

Vol. 4 Nataly Tcherepashenets (ed.)
Globalizing On-line: Telecollaboration, Internationalization, and Social Justice.
2015. ISBN 978-3-0343-1520-3

www.ingramcontent.com/pod-product-compliance
Ingram Content Group UK Ltd.
Pitfield, Milton Keynes, MK11 3LW, UK
UKHW021851100926
8556IPUK00020B/1177

9 783034 315203